IMPORTANT BIRD AREAS OF

JAMMU & KASHMIR

PRIORITY SITES FOR CONSERVATION

IMPORTANT BIRD AREAS OF
JAMMU & KASHMIR
PRIORITY SITES FOR CONSERVATION

Asad R. Rahmani, Zafar-ul Islam, Khursheed Ahmad,

Intesar Suhail, Pankaj Chandan, and Ashfaq Ahmed Zarri

Maps prepared by

Mohit Kalra

Layout and design by

V. Gopi Naidu

Supported by

Oxford University Press, Walton Street, Oxford OX2 6DP
Oxford, New York,
Athens, Auckland, Bangkok,
Cape Town, Chennai, Dar-es-Salaam,
Delhi, Florence, Hong Kong, Istanbul,
Karachi, Kolkata, Kuala Lumpur, Madrid, Melbourne,
Mexico City, Mumbai, Nairobi, Paris,
Singapore, Taipei, Tokyo, Toronto,
and associated companies in
Berlin, Ibadan

Recommended citation:
Rahmani, A.R., Zafar-ul Islam, Khursheed Ahmad, Intesar Suhail, Pankaj Chandan, and Ashfaq Ahmed Zarri (2012) *Important Bird Areas of Jammu & Kashmir*
Indian Bird Conservation Network, Bombay Natural History Society, Royal Society for the Protection of Birds and BirdLife International. Oxford University Press. Pp. xii + 152.

Consultant Editor: Gayatri W. Ugra
Layout and design: V. Gopi Naidu
Maps: Mohit Kalra

© IBCN: Bombay Natural History Society, 2012
IBCN, c/o BNHS, Hornbill House, Shaheed Bhagat Singh Road, Mumbai — 400 001, India
Telephone: 0091-22-22821811, Fax: 0091-22-22837615
Email: ibabnhs@gmail.com, bnhs@bom4.vsnl.net.in
Websites: <www.ibcnetwork.org> <www.bnhs.org>

Bombay Natural History Society is registered in India under the Bombay Public Trust Act 1950: F244 (Bom) dated July 6, 1953

ISBN : 978-0-19-809218-6

Proceeds from the sale of this book will go to the Indian Bird Conservation Network

Front Cover: Himalayan Monal by Nikhil Devasar

Back Cover: White-backed Vulture by Dhritiman Mukherjee

Available from
IBCN, c/o BNHS, Hornbill House, Shaheed Bhagat Singh Road, Mumbai — 400 001, India
Telephone: 0091-22-22821811, Fax: 0091-22-22837615
Email: ibabnhs@gmail.com, bnhs@bom4.vsnl.net.in
Websites: <www.ibcnetwork.org> <www.bnhs.org>

Processed by TRENDZ PHOTOTYPESETTERS. Email: gotrendz@gmail.com
Printed by SPECIFIC ASSIGNMENTS INDIA PVT. LTD. Email: info@specificassignments.com

CONTENTS

PREFACE

Important Bird Areas (IBAs) is one of the major programmes of BirdLife International and its Partners in more than 100 countries. Till now nearly 12,000 IBAs have been identified worldwide, to which India has contributed 466 IBAs, perhaps the largest number in any country. Though started by NGOs and civil society, IBAs are now increasingly being recognized by various governments as sites of high biodiversity importance, or Key Biodiversity Areas (KBAs). Birds are considered good indicators of good biodiversity sites. As the name suggests, IBAs are identified based on birds as the main criteria, but almost all IBAs are good sites for protection of other taxa.

Many protected areas in India (and probably elsewhere) were not selected or prioritised according to biodiversity criteria (except for tiger reserves), which is why some very high biodiversity areas have been left out, while not so important areas or areas with huge human presence have been included in the Protected Area system. The IBA process is rigorous and scientific, and only those areas get selected that fulfill IBA criteria. The selection of IBAs is also a dynamic process and new IBAs are added, while some are deleted, if due to any reason they fail to fulfill IBA criteria any longer.

In 2004, the Bombay Natural History Society and BirdLife International, with the support of NGOs, governments and members of civil society, identified 466 IBAs in India, including 21 in Jammu & Kashmir. As the main IBA inventory is a large, bulky (1133 pages) and costly (Rs. 3,000) volume, it is not easily accessible to decision makers, researchers, and students. As conservation action generally takes place at state level, it was felt that additional state-level IBA books would be more useful. The first such book was brought out by Sikkim, the second by Uttar Pradesh, and this is the third one. One more, on the Important Bird Areas of Maharashtra, is under production.

Besides describing the existing 21 IBAs of Jammu & Kashmir, this book also suggests seven new IBAs. We are sure there are more sites that may qualify to be IBAs, but we need more information on such sites and their avifaunal diversity. The site accounts have been corrected and updated. The main IBA book did not have polygonal maps, which have now been done. New pictures of birds and habitats have been included.

With its low price and easily accessibility in the State, it is hoped that this book will be used by forest officers, decision makers, researchers and birdwatchers for the protection of important biodiversity sites. With increasing interest in birdwatching, IBA books are good guides for travellers. If better protection is given to birds and other wildlife, and their habitats, the purpose of this book will be served.

Authors

GOVERNOR
JAMMU & KASHMIR

MESSAGE

I am happy to learn that the Bombay Natural History Society (BNHS) and World Wide Fund for Nature have collaborated to bring out an updated volume on the "Important Bird Areas of Jammu & Kashmir".

The State of Jammu & Kashmir is endowed with unmatched natural beauty and extremely rich biodiversity. J&K has beautiful rivers, lakes and many wetlands which attract waterfowl from different parts of the State and from distant lands during winter. To protect its flora and fauna the State Government have notified certain areas as "protected" reserves, among which perhaps the most well known is the Dachigam National Park.

BNHS, the oldest conservation organization in India, has the distinction of working in partnership with BirdLife International (BLI) which has the objective of identifying Important Bird Areas (IBAs) all over the world. After working together for several years BNHS and BLI have been able to identify 466 IBAs in the country, of which as many as 21 are located in our State.

I thank BNHS for bringing out a volume on the IBAs in J&K. I believe that the State Wildlife Department shall be able to identify many more IBAs in the State if it works in close collaboration with conservationists and the universities, particularly the two State Farm Universities.

On the basis of my long experience in the field, I feel that the State Wildlife Department would be able to protect the birdlife in J&K if it works in close coordination with the local populations in the various IBAs.

I am sure that the forthcoming book on the IBAs of Jammu & Kashmir will be a precious addition to the existing literature on the biodiversity of our State.

10th September, 2012
Srinagar

(N. N. Vohra)

MESSAGE

I am delighted to know that the Bombay Natural History Society (BNHS), Mumbai, one of the oldest conservation NGOs of the world, is going to publish a book, 'Important Bird Areas of Jammu & Kashmir'.

Human forays into natural ecosystems, urbanisation, population growth and expansion of activities into the domain of nature have disturbed the natural world, floral and faunal ecosystems. These have far reaching and wide implications on biodiversity, flora, fauna, and natural systems. The implications are largely negative. Given this, measures need to be taken that preserve the natural biodiversity of our world.

In this schema, the Bombay Natural History Society's book titled 'Important Bird Areas of Jammu & Kashmir' is a timely, topical, and germane academic effort to preserve, conserve, and protect biodiversity. The painstaking research conducted by the authors will surely help in conservation and protection of Important Bird Areas (IBAs) — an effective way of identifying conservation priorities — in the state of Jammu & Kashmir.

Given that a site is recognised as an IBA only if it meets certain criteria, based on the occurrence of key bird species that are vulnerable to global extinction, or whose populations are vulnerable to global extinction, and birds are shown to be an effective indicator of biodiversity, the book's recommendations would constitute an immense contribution. Its detailed accounts of the conservation status of the floral and faunal diversity with special reference to the birdlife of the 21 Important Bird Areas (IBAs) and seven potential IBAs would be invaluable.

The cases in point are the wetlands of J&K — Dal and Wular Lakes, Hokarsar, Haigam, Shallabugh, Mirgund Wetland Reserves in the Kashmir Valley, and Tso Moriri, Tso Kar, Chushul, and Hanle Marshes in the Ladakh region — besides supporting many species of wetland plants and insects provide a critical habitat for ducks, geese, and many other waterfowl species. These can potentially give a fillip to ecotourism in the Valley.

In sum, the book will help the both the State Forest and Wildlife Protection Departments with improved scientific-based management and conservation of birds and their habitats, with a focus on Important Bird Areas of the State.

I convey my best wishes to the organization for its sterling efforts in publishing the book.

(Omar Abdullah)

MESSAGE

Identification of Important Bird Areas (IBAs) is a very successful programme of BirdLife International, UK and its partners in more than 115 countries. In India, Bombay Natural History Society (BNHS) is a key BirdLife Partner. The BNHS works with several national and regional non-government organizations, and WWF-India is one of them. Even during the identification of IBAs in India about 10 years ago, WWF-India and its staff had played a major role. The inventory of IBAs of India was published in 2004, listing 466 IBAs. Since its publication, many more areas have been identified as IBAs, mainly through the support of non-government organizations and government officials.

The present book, IMPORTANT BIRD AREAS OF JAMMU & KASHMIR, is an exemplary collaboration between two of India's larger conservation organizations, the BNHS and WWF-India. Besides describing and updating information on the existing 21 IBAs that were identified in 2004, seven potential Important Bird Areas have been described in brief with the help of new information.

The book is profusely illustrated with the habitats of different IBAs and the birds found therein. Boundary maps have been included, which further enhance the value of this book, particularly for researchers and decision makers. While research needs to continue to add to our knowledge in the region, I hope this book will stimulate greater interest in the avifauna of Jammu and Kashmir that is known for its natural beauty and protected area network. Many IBAs are within protected areas, but there are many more that need immediate protection for their long-term survival.

WWF-India is honoured to partner in this project, as funding sponsors and contributors.

Ravi Singh,
CEO, WWF-India

ACKNOWLEDGEMENTS

First of all, we would like to acknowledge the support of Mr. Ravi Singh, SG & CEO, WWF-India, Dr. Sejal Worah, Programme Director, and Dr. Parikshit Gautam, Director, Branches and Special Projects, WWF-India for sponsoring this book.

Our sincere gratitute to Mr. A.K. Srivastava, IFS, IG, MoEF, Government of India, and former Chief Wildlife Warden of Jammu and Kashmir; Mr. Jigmet Takpa, IFS, Conservator Forest (Wildlife) Ladakh Region, Department of Wildlife Protection, Leh, Ladakh; Mr. Asif M. Sagar, IFS, Conservator Forest (Wildlife) Jammu Region, Department of Wildlife Protection, Jammu.

Ms Archana Chatterjee, National Project Coordinator, UNESCO, New Delhi; Ms Yamini Panchaksharam, Sr. Programme Officer, WWF-India; Mr. Kishor Chandra, Administration Officer, WWF-India; Ms Nisa Khatoon, Senior Project Officer, WWF-India Field Office, Leh; Mr. Pushpinder Singh Jamwal, Project Officer, WWF-India, Field Station Gharana, Jammu; Mr. Rohit Rattan, Project Officer, WWF-India, Field Office, Pir Panjal Range, Jammu; Mr. Shakeel Ahmed, Field Assistant, WWF-India, Field Office, Pir Panjal Range, Jammu; Mr. Phuntsog Tashi, Project Officer, WWF-India Field Office, Tso Moriri, Ladakh; Mr. Tsewang Rigzin, Assistant Project Officer, WWF-India Field Office, Tso Kar, Ladakh; Mr. Dawa Tsering, Field Assistant, WWF-India Field Office, Tso Moriri, Ladakh; Mr. Mohd Kazim, Field Assistant, WWF-India Field Office, Kargil, Ladakh; Mr. Abdul Rauf, Wildlife Warden (Leh), Department of Wildlife Protection, Leh, Ladakh; Mr. Tsering Angchuk, Wildlife Warden (Kargil), Department of Wildlife Protection, Leh, Ladakh; Mr. Tsering Angchuk, Range Officer, Leh, Department of Wildlife Protection, Leh, Ladakh; Mr. Lobzang Khatup, Range Officer, Nubra, Department of Wildlife Protection, Leh, Ladakh; Mr. Tsering Phunchuk, Range Officer, Changthang, Department of Wildlife Protection, Leh, Ladakh; Mr. Mohd Ali, Divisional Forest Officer, Forest Department, Kargil; Mr. Mohd Abbas, Range Officer, Forest Department, Kargil.

Ms Radhika Kothari, Deputy Director, Snow Leopard Conservancy India Trust; Mr. Jigmet Dadul, Programme Manager (Conservation & Livelihood), Snow Leopard Conservancy India Trust; Ms Tsering Angmo, Programme Manager (Education & Outreach), Snow Leopard Conservancy India Trust; Ms Rigzin Chorol, General Manager (Administration & Finance), Snow Leopard Conservancy India Trust; Ms Tsering Lazes, Office Assistant, Snow Leopard Conservancy India Trust; Mr. Zafar Khan, Range Officer, Wildlife, Department of Wildlife Protection, Rajouri, J&K; Dr. R.B. Srivastava, Scientist G, Director, Defence Institute of High Altitude Research, Leh, Ladakh; Dr. O.P. Chourasia, Scientist 'F', Dy Director, Defence Institute of High Altitude Research, Leh; Dr. Tsering Stobdan, Scientist 'E', Defence Institute of High Altitude Research, Leh.

From Sher-e-Kashmir University of Agricultural Sciences & Technology (SKUAST) we wish to thank Professor Anwar Alam, former Vice Chancellor; Dr. Tej Partap, Vice Chancellor; Dr. A.R. Trag, former Director Research and present Vice Chancellor, Islamic University of Science & Technology, Awantipora, Kashmir; Dr. G.M. Wani, former Director Extension Education; Dr. Syed Sajjad Hussain, former Director Resident Instructions (DRI); Dr. M.A. Kirmani, former Dean, Faculty of Veterinary Sciences & Animal Husbandry; Dr. Shafiq A. Wani, Director Research; Dr. M.A. Gora, former Registrar; Dr. F.A. Zaki, Registrar; Mr. Parvez Ahmad Bhat, Secretary to the Vice Chancellor, Mr. Mubashir Rashid Qadri, AEE (R & A), Dr. M.T. Banday, Professor & Head, Division of LPM.

The following officials have also been of tremendous help to us, namely Dr. Mehmood ur Rahman, IAS (Retd.), Former Secretary to Government of India, Parliamentary Affairs and former Vice Chancellor, AMU, Aligarh; Mr. Iqbal Khanday, IAS, Principal Secretary to Government of J&K, Finance Department; Mr. Shantanu, IAS, former Commissioner Secretary to Government of J&K Forest Department; Professor G.M. Untoo, former Principal, Government Islamia College, Srinagar; and Dr. C.M. Seth, former Chief Wildlife Warden, J&K Government, Director WWF-India, J&K.

Mr. A.R. Wani, IFS (Retd.), former PCCF & CWLW, J&K Government, needs special acknowledgement from us as he has always supported us at every stage.

From the Department of Wildlife Protection, J&K, we would like to acknowledge the following persons:

Mr. A.K. Singh, IFS, Chief Wildlife Warden, J&K; Mr. M.A Tak, IFS, Regional Wildlife Warden, Kashmir; Mr. A.R. Wadoo, IFS (Rtd.), former Chief Wildlife Warden, J&K; Mr. O.P. Sharma, IFS, Mr. S.F.A Gillani, IFS, and Mr. Naseer A. Kichloo, IFS. We also want to thank Dr. Mir Mansoor, Mr. M.A. Parsa, Mr. Rashid Y. Naqash, Mr. A.B. Rauf Zargar, Mr. Tahir Shawl, Mr. Imtiyaz Lone, Ms Ifshan Dewan, Mr. M.M. Baba, Mr. G.A. Lone, Mr. M. Sadiq Mir, Ms Sameena Amin, Mr. Junaid Nazir Shah, Mr. Shahid Bashir Khan, Mr. Riyaz Ahmad, Mr. Mohammad Raashid, Mr. Tassaduq Moeen, Mr. Athar Noor, Mr. Mudasir Mansoor and Mr. Mansoor Nabi Sofi.

From BirdLife, we would like to thank Dr. Marco Lambertini, Dr. Nigel Collar, Dr. Mike Crosby, Dr. Richard Grimmett, Dr. A.J. Stattersfield; and from RSPB we would like to thank Dr. Mike Clarke, Dr. Tim Stowe, Dr. Dieter Hoffman and Mr. Ian Barber.

Our sincere gratitude to Mr. Homi Khusrokhan, President, BNHS, Mrs. Usha Thorat, Vice President, Dr. Ravi Chellam, Vice President, Ms Sumaira Abdulali, Hon. Secretary , Mr. E.A. Kshirsagar, Hon. Treasurer, and all members of the Governing Council of BNHS for their support. We would also like to thank Dr. Ashok S. Kothari, former Hon. Secretary of BNHS.

Among the staff of BNHS, we would like to thank the following : Mr. Abhijit A. Malekar, Mr. Ajit Majgaonkar, Mr. Anil Malhotra, Mr. Amjad Hussain, Mr. Asif N. Khan, Mr. Atul Sathe, Ms Darshana Patil, Ms Deepa Fernandes, Mr. Deepak Apte, Ms Divya Varier, Mr. Divyesh

Parikh, Ms Gisha Shankar, Mr. Isaac Kehimkar, Mr. J.P.K Menon, Mr. M.G. Mathews, Ms Mirium Abraham, Ms Neha Sinha, Ms Nikita V. Prakash, Ms Nirmala Reddy, Dr. Parag Deori, Dr. Raju Kasambe, Mr. Sachin Kulkarni, Mr. Sujit Narwade, Dr. V. Shubhalaxmi, Ms Varsha Chalke, Dr. Vibhu Prakash and Ms Vibhuti Dedhia. Special thanks to Dr. Gayatri W. Ugra for editing this book.

Last but not the least we want to thank the following photographers for providing their images for this book: Mr. Anant Zanjale, Mr. Asif Khan, Mr. Clement Francis, Mr. Dawa Tsering, Ms Garima Bhatia, Mr. Gaurav Sharma, Ms Nisa Khatoon, Mr. Otto Pfister, Mr. Pankaj Chandan, Mr. Phuntsog Tashi, Mr. Rajat Bhargava, Mr. T.K. Sajeev, Mr. Sachin Rai, Mr. Siddharth Pandey, Thakur Dalip Singh, Mr. Trevor Price, Mr. Vivek Sinha and Mr. Yeshey Dorji.

INTRODUCTION

From ice-clad peaks to subtropical valleys and permanent wetlands, the habitats of Jammu & Kashmir state are as varied as its biodiversity and avifauna

India is one of the twelve megadiversity countries in the world, and divided into 10 biogeographical regions based on the landmass and species distribution. The state of Jammu & Kashmir falls under the Western Himalayan and Trans-Himalayan biogeographical region of India, and is at the intersection between the temperate Palaearctic and tropical Oriental biogeographic regions of the world. Owing to the enormous diversity of habitat types, great altitudinal span of the mountains, and climatic variations, Jammu & Kashmir encompasses a diversity of plant and animal species, many of which are endemic or near-endemic to this area.

Jammu & Kashmir (32° 17' 37° 05' N and 72° 31' 80° 20' E) is bounded on the north by China (Karakoram range), on the east by Tibet, on the west by Pakistan and Afghanistan, and to the south by Himachal Pradesh and Punjab. This hilly state is divided into three geographical regions, namely, the temperate valley and mountains of Kashmir, the cold desert of Ladakh and the subtropical plains of Jammu. The higher regions are covered by Pir Panjal, Karakoram, and the inner Himalayan ranges. The climate varies from subtropical in the Jammu region to cold and arid in Ladakh. The state has a geographical area of 22.22 million ha (6.8% of India's geographical area). It is the only state in India with one capital in summer and another in winter. Srinagar is the capital city in summer and Jammu is the winter capital.

IBAs of Jammu & Kashmir

IBA site codes	IBA site names	IBA criteria
IN-JK-01	Chushul Marshes	A1
IN-JK-02	Dachigam National Park	A1, A2, A3
IN-JK-03	Dehra Gali (DKG) Forest	A1, A2
IN-JK-04	Gulmarg Wildlife Sanctuary	A1, A2, A3
IN-JK-05	Haigam Rakh (Wetland)	A1, A4iii
IN-JK-06	Hanle Plains/Marshes	A1
IN-JK-07	Hemis High Altitude National Park	A3
IN-JK-08	Hirpora Wildlife Sanctuary	A1, A2
IN-JK-09	Hokarsar Wetland Conservation Reserve	A1, A4iii
IN-JK-10	Kishtwar National Park	A1, A2, A3
IN-JK-11	Lachipora Wildlife Sanctuary	A1, A2
IN-JK-12	Limber Valley Wildlife Sanctuary	A1, A2, A3
IN-JK-13	Mirgund Jheel and Reserve	A1, A4i
IN-JK-14	Overa-Aru Wildlife Sanctuary	A1, A2, A3
IN-JK-15	Pangong Tso	A1, A3
IN-JK-16	Ramnagar Wildlife Sanctuary	A1
IN-JK-17	Shallabugh Conservation Reserve	A4iii
IN-JK-18	Tso Kar Basin	A1
IN-JK-19	Tso Moriri Lake and Adjacent Marshes	A1, A4i
IN-JK-20	Wular Lake and Associated Marshes	A1, A4iii
IN-JK-21	Gharana Wetland Reserve	A4iii

IMPORTANT BIRD AREAS PROGRAMME

The IBA Programme of BirdLife International aims to identify, monitor and protect a global network of Important Bird Areas (IBAs) for the conservation of the world's birds and other biodiversity. BirdLife Partners take responsibility for the IBA Programme nationally, with the BirdLife Secretariat taking the lead on international aspects and in some priority non-Partner countries. As of 2012, nearly 12,000 sites in some 200 countries and territories have been identified as Important Bird Areas. In India, BirdLife International and Bombay Natural History Society have identified 466 IBAs. In Jammu & Kashmir, there are 21 IBAs, with many more areas as potential IBAs.

The selection of Important Bird Areas has been a particularly effective way of identifying conservation priorities. IBAs are key sites for conservation – small enough to be conserved in their entirety and often already part of a protected area network. They do one (or more) of three things:

(1)   Hold significant numbers of one or more globally threatened species,

(2)   Are one of a set of sites that together hold a suite of restricted-range species or biome-restricted species, and

(3)   Have exceptionally large numbers of migratory or congregatory species.

For details, see www.birdlife.org

Important Bird Areas of Jammu & Kashmir

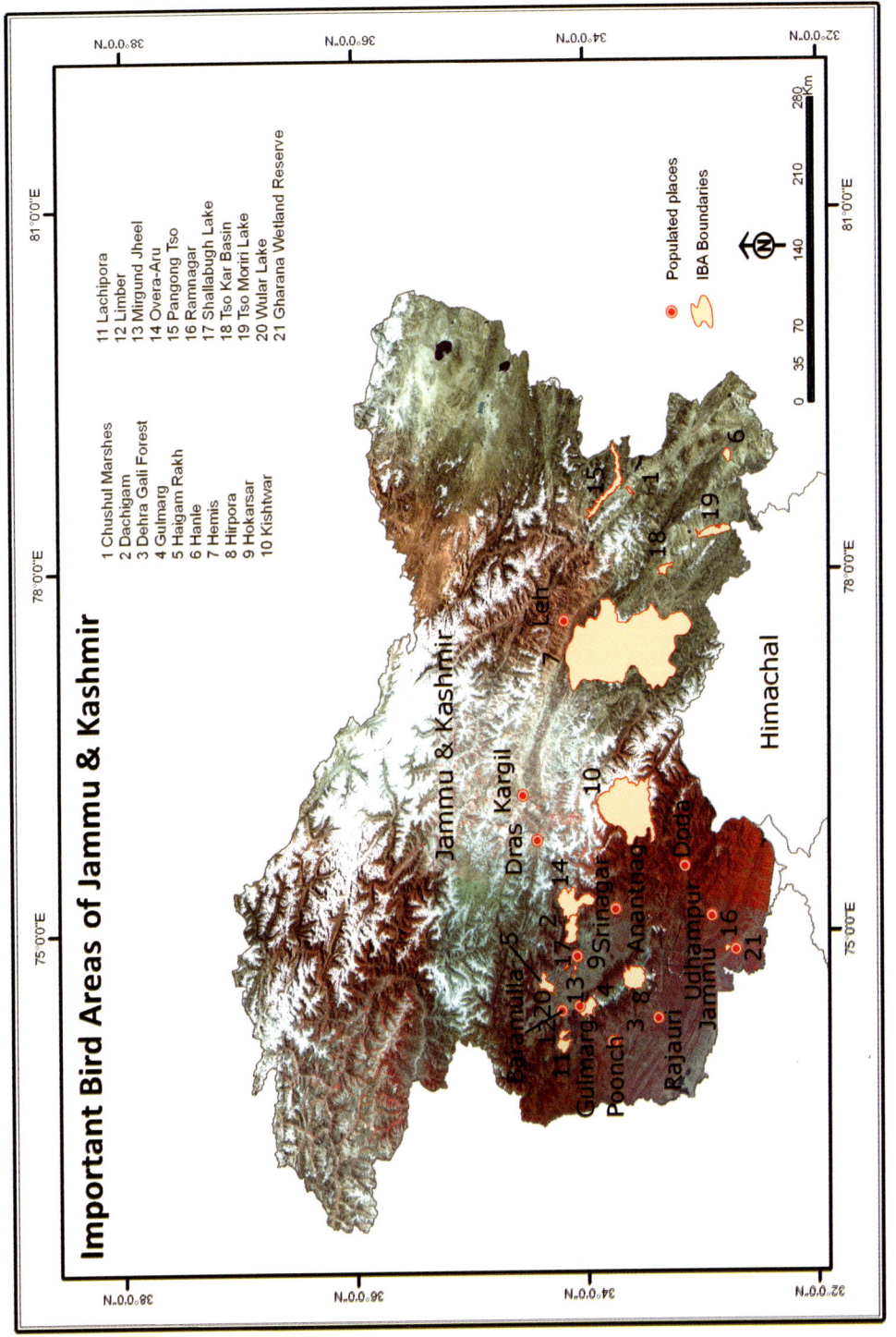

1 Chushul Marshes
2 Dachigam
3 Dehra Gali Forest
4 Gulmarg
5 Haigam Rakh
6 Hanle
7 Hemis
8 Hirpora
9 Hokarsar
10 Kishtwar

11 Lachipora
12 Limber
13 Mirgund Jheel
14 Overa-Aru
15 Pangong Tso
16 Rannagar
17 Shallabugh Lake
18 Tso Kar Basin
19 Tso Moriri Lake
20 Wular Lake
21 Gharana Wetland Reserve

Populated places
IBA Boundaries

0   35   70   140   210   280
Km

Jammu & Kashmir

Himachal

Leh

Dras   Kargil

Srinagar   Anantnag   Doda

Baramulla   Gulmarg   Poonch   Rajauri   Udhampur   Jammu

From ice-clad mountains to subtropical forests, the state has a vast range of habitats

ASAD R. RAHMANI

Why IBAs are important

Important Bird Areas (IBAs) are sites of international significance for the conservation of birds and their habitats at the global, regional and sub-regional level. The selection of IBAs is a particularly effective way of identifying conservation priorities. IBAs are key sites for conservation – small enough to be conserved in their entirety and often already part of a protected area network.

A site is recognised as an IBA only if it meets certain criteria based on the occurrence of key bird species that are vulnerable to global extinction or whose populations are otherwise irreplaceable.

An IBA must be amenable to conservation action and management. The IBA criteria which are applicable globally are as follows:

A1: Sites holding globally threatened bird species of global conservation concern.

A2: Sites having restricted-range bird species, i.e. bird species with a historic breeding range up to 50,000 sq. km in the world.

A3: Sites having biome-restricted bird species, i.e. bird species representing the distinct habitat types.

A4: Sites having large congregations of birds.

What is the significance of IBAs?

FOR CONSERVATION

- IBAs help identify priority sites for conservation action.
- IBAs provide the framework to monitor and manage sites of global conservation significance.
- IBAs provide decision makers with high quality information needed to formulate a national conservation strategy and implement international agreements.
- IBA programme helps develop national and local capacity for biodiversity conservation.

FOR COMMUNITIES

- IBAs help meet daily subsistence needs of communities for food, fuel, fodder, and other natural resources.
- IBAs are a source of livelihood for many communities who harvest minor forest produce for sale in local markets.
- IBAs are a part of distinct indigenous cultures and a repository of traditional knowledge resources.

FOR CLIMATE CHANGE

- IBAs play an important biological role as carbon sinks, thereby reducing the amount of CO_2 in the atmosphere.
- IBAs help mitigate the impact of extreme weather events such as drought, flash floods and cyclones by acting as a buffer for human habitations.
- IBAs help climate change affected communities to cope by providing water, food, and building material for temporary shelters.

Despite recent strife, Jammu & Kashmir is one of the favourite tourist destinations in India. Most tourists go there to enjoy scenic beauty. Considering its high biodiversity and well managed protected areas, the state can become a major attraction for birdwatchers

Agriculture is the mainstay of the State's economy. Rice *Oryza sativa*, wheat *Triticum aestivum*, and maize *Zea mays* are the major crops. Barley *Hordeum vulgare*, bajra *Pennisetum glaucum*, and jowar *Sorghum vulgare* are cultivated in some parts. Kashmir handicrafts have always been a byword for excellence. This sector provides employment to about 0.2 million people. Kashmir carpets earn substantial foreign exchange (Mathew 2003). The 300 km Srinagar-Jammu National Highway and the 265 km Mughal Road via Pir Panjal mountains are the two major surface links between the Kashmir Valley and the rest of the country. Kashmir is internationally known for its beauty and is a favourite tourist destination. The main tourist centres are Srinagar, Pahalgam, Gulmarg, and Sonamarg. Hindu pilgrim centres of special importance include Amarnath and Vaishno Devi. Ladakh has also become a major tourist destination, bringing problems of over-crowding, litter, and uncontrolled mushrooming of hotels.

The total human population of the State according to the 2001 Census was 10.07 million, which was about 1.0% of the country's population. Nearly 75% of the people lived in the villages. The population density was 45 persons per sq. km, which is very low compared with other states. This is due to the large uninhabited cold desert area in Ladakh. The human population increased to 12,548,000 in 2010, an increase of about 23% in 10 years (Population Census 2010).

The Tibetan Plateau and Pir Panjal Range are a source of many major rivers, such as the Indus and the Jhelum, which start from this region, but much of the water is drained internally, where the rivers end in vast lakes and marshes. These inland wetlands,

VIVEK SINHA

particularly marshes and peat bogs, absorb and hold huge quantities of fresh water which they release slowly during both dry and wet seasons, preventing floods and providing sufficient water for the sustenance of millions of people. The wetlands of J&K include the Dal and Wular Lakes, Hokarsar, Haigam, Shallabugh, and Mirgund Wetland Reserves in the Kashmir Valley; Pangong Tso, Tso Moriri, Tso Kar, Chushul, and Hanle marshes in the Ladakh region; and Gharana in Jammu region. Besides supporting many species of wetland plants, animals, and insects, these wetlands provide critical habitats for ducks, geese, swans, and many other waterfowl species. Such lakes and marshes, mostly freshwater and saline respectively, are important as wintering and/or breeding grounds for a diversity of waterfowl such as the Vulnerable Black-necked Crane *Grus nigricollis*, as well as Greylag Goose *Anser anser*, Bar-headed Goose *Anser indicus*, Great Crested Grebe *Podiceps cristatus*, Mallard *Anas platyrhynchos*, Ferruginous Duck *Aythya nyroca*, and Blunt-winged Warbler *Acrocephalus concinens*, besides a variety of other waterfowl.

The wetlands of Kashmir Valley are important for both resident and migratory waterfowl. They are major wintering areas for a variety of migratory ducks and geese, and extremely important breeding areas for some species. They are particularly important for long distance migrants, as a stopover site for feeding and resting. Many waterbirds occur in huge numbers in the wetlands of Jammu & Kashmir, much above the proportion of the total species abundance (1%) determined by Wetlands International (2006) as one of the criteria for declaring a wetland as a Ramsar Site.

TREVOR PRICE

Livestock overgrazing is one of the major problems, particularly in fragile high altitude areas

Wetlands of Kashmir Valley are extremely important for biodiversity and people. However, most of the wetlands are suffering from encroachment, pollution, drainage, and plantation. Even an important wetland like Wular (above) which is a Ramsar Site has been converted by extensive plantation of poplar, which the government now wants to remove

Vegetation

Broadly, Jammu & Kashmir has five types of vegetation, namely Subtropical Dry Evergreen, Himalayan Moist Temperate, Himalayan Dry Temperate, Subalpine and Alpine Forests. The recorded forest area is 2.02 million ha, which constitutes 9% of the geographical area of the State. Forests are largely distributed in the Kashmir Valley and Jammu region. Dense forest and open forest account for 1,184,800 ha and 938,900 ha respectively, according to the Ministry of Environment and Forests (2001). There are 22 districts in the State. The western districts have more forest cover with dense and open forests. Reasi, Poonch, Kathua, and Jammu have more forest cover than Ladakh, Gilgit, Baramulla, and Anantnag (Ministry of Environment and Forests 2001). Ladakh region, being part of Trans-Himalaya, is devoid of forest vegetation, although this area is now being extensively planted with Willow *Salix* spp., which is affecting bird life. Ladakh region is a wellknown cold desert with highly adapted flora and fauna (Gujja *et al.* 2003).

IBAs and Protected Areas

Jammu & Kashmir has been the pioneer state in the field of conservation and has had a network of wildlife protected areas (Game Reserves) from the time of the erstwhile Maharajas of the state. These game reserves were covered by the former Game

Preservation Act, 1852 which was revised and updated as the J&K Wildlife Protection Act, 1978 (Amended 2002). Some of the sanctuaries were established nearly one hundred years ago, mainly to protect the catchment of important lakes and to provide good hunting for the Maharaja. Since then the State Government has notified about 16,000 sq. km under the Protected Area Network (PAN) which is 15.6% of the total geographical area of the State, comprising five National Parks, 14 Wildlife Sanctuaries, and 35 Conservation Reserves. The Protected Areas (PAs) include 2,762 sq. km of forest (12.76% of 20,230 sq. km total forest area). The remaining 13,150 sq. km is the high altitude cold desert area of Ladakh. It is remarkable that at the national level, 4.8% of the total geographical area is under the PAN, whereas percentage coverage of PAs in J&K is nearly 3 times more than the national average. J&K has also got the largest number of PAs in the country, i.e., 53. In terms of the geographical coverage of the PAs also, the State is second only to Gujarat.

BNHS and BirdLife International have identified 21 Important Bird Areas (Islam & Rahmani 2004).

Of the Protected Areas, Dachigam National Park is of special ecological significance as it harbours the last viable population of the globally Threatened Kashmir Red Deer or Hangul *Cervus elaphus hanglu*. Wular Lake, situated in Baramulla district, comprising an area of 8,900 ha, is a wetland of international importance which was declared as one of the first six Ramsar Sites in India in 1990. Of the four national parks, three have been identified as IBAs. Of the 16 wildlife sanctuaries, eight are IBAs. Of the 21 Important Bird Areas (IBAs) identified in Jammu & Kashmir in 2004, 11 fulfill Ramsar criteria (Islam & Rahmani 2008).

A large part of the State is covered with Himalayan Temperate Forest

Potential IBAs

Since the publication of *Important Bird Areas in India* (Islam & Rahmani 2004), the following 7 areas have been identified as potential IBAs:

1. Aharbal-Kounsarnag Forests
2. Gurez Valley
3. Kanji Wildlife Sanctuary
4. Rangdum Wetlands
5. Sheshera Reserve Forest
6. Shikargah Conservation Reserve, Tral
7. Suru Valley

The description of potential IBAs is given at the end of the book.

AVIFAUNA

Jammu & Kashmir lies in the Western Himalaya Endemic Bird Area (EBA 128) where 11 restricted-range species have been listed by Stattersfield *et al.* (1998). Because of great altitudinal variation and differing physiogeographical regions, Jammu & Kashmir has three biomes: Biome 5 (Eurasian High Montane – Alpine and Tibetan) above 3,600 msl; Biome 7 (Sino-Himalayan Temperate Forest) mainly *c.* 1,800 to 3,600 msl; and Biome 8 (Sino-Himalayan Subtropical Forest) between *c.* 1,000 and 2,000 msl. The Eurasian High Montane (Alpine and Tibetan) Biome is mainly distributed in the Ladakh region, especially in Changthang plateau. The Sino-Himalayan Temperate Forest type habitat is present in most of the IBAs in the State.

In the Kashmir Valley, many protected areas support restricted-range species and some water bodies support large congregations of migratory waterbirds. These restricted-range species occur mainly in Temperate Coniferous or Broadleaf Forest, Subalpine Forest, and Montane Grasslands. For example, the Kashmir Flycatcher *Ficedula subrubra*, which is considered a globally Threatened species, is found between 1,800 and 2,700 msl in Temperate Mixed Broadleaf Forest, especially where there is dense growth of Parrotia *Parrotiopsis jacquemontiana* (Stattersfield *et al.* 1998). Other similar species, namely Tytler's Leaf-warbler *Phylloscopus tytleri* and White-throated Tit *Aegithalos niveogularis* are found between 1,500 and 3,600 msl in Pine, Oak, and Mixed Deciduous Forests. Other restricted-range species which are found in or near the Valley are the Kashmir Nuthatch *Sitta cashmirensis*, Spectacled Finch *Callacanthis burtoni*, and Orange Bullfinch *Pyrrhula aurantiaca*. The two finches are found in open Coniferous Forest, Mixed Woodland Forest, Deciduous Forest, and occasionally in stands of Birch (Stattersfield *et al.* 1998).

The Changthang region in Ladakh is an important breeding ground for waterbirds. Apart from hosting the largest breeding congregation of Bar-headed Goose *Anser indicus* in India, the Changthang region also supports the largest population of the globally Vulnerable Black-necked Crane *Grus nigricollis* in India (Chandan *et al.* 2008a). During a study on its breeding ecology, Pfister (1998) recorded 12 sites in the Changthang region as breeding sites for this Vulnerable species and counted 38 Black-necked Cranes. In a subsequent survey of Changthang in 2001, 42 individuals were counted, with 10 breeding pairs in the Changthang region (S.A. Hussain, *pers. comm.* 2003). During subsequent

Western Tragopan	*Tragopan melanocephalus*	IN-JK-03, 10, 11, 12
Cheer Pheasant	*Catreus wallichii*	IN-JK-12, 10
Tytler's Leaf-warbler	*Phylloscopus tytleri*	IN-JK-02, 14
Kashmir Flycatcher	*Ficedula subrubra*	IN-JK-02, 03, 04, 08, 14
Kashmir Nuthatch	*Sitta cashmirensis*	IN-JK-14, 02
Orange Bullfinch	*Pyrrhula aurantiaca*	IN-JK-02, 14

years, a study on the status and ecology of the Black-necked Crane was conducted by WWF-India. Till 2011, six new nesting sites were recorded (Chandan *et al.* 2006). During the 2011 WWF survey, a population of 81 Black-necked Crane was recorded in Ladakh.

Another IBA in Ladakh is the Hemis National Park, which is important for all the high altitude birds of the Western Himalaya. About 80 bird species are found in the Park and 50 of them breed there. The important bird species recorded by Khursheed (1999, *pers. comm.* 2012) in Markha Valley of Hemis National Park include Bearded Vulture or Lammergeier *Gypaetus barbatus*, Red-billed Chough *Pyrrhocorax pyrrhocorax*, Yellow-billed Chough *Pyrrhocorax graculus*, Robin Accentor *Prunella rubeculoides*, Common Great Rosefinch *Carpodacus rubicilla*, Streaked Great Rosefinch *C. rubicilloides*, Black-throated Thrush *Turdus ruficollis*, and Brown Accentor *Prunella fulvescens*.

Changthang in Ladakh is a huge area and many wetlands and other important spots (e.g., Sumdo near Puga) are included in this IBA. There are many small wetland sites which are important breeding grounds for waterbirds but do not fulfill IBA criteria, so the Changthang plateau as a whole could be considered one IBA. This does not mean

Kashmir Flycatcher *Ficedula subrubra* is found in five to six IBAs

Black-necked Crane *Grus nigricollis*

YESHEY DORJI

List of threatened birds of Jammu & Kashmir

CRITICALLY ENDANGERED

White-backed Vulture	*Gyps bengalensis*	IN-JK-16
Slender-billed Vulture	*Gyps tenuirostris*	IN-JK-16

ENDANGERED

Egyptian Vulture	*Neophron percnopterus*	IN-JK-02, 08

VULNERABLE

Marbled Teal	*Marmaronetta angustirostris*	IN-JK-20
Pallas's Fish-eagle	*Haliaeetus leucoryphus*	IN-JK-05, 09, 20
Greater Spotted Eagle	*Aquila clanga*	IN-JK-06
Eastern Imperial Eagle	*Aquila heliaca*	IN-JK-02
Western Tragopan	*Tragopan melanocephalus*	IN-JK-03, 10, 11, 12
Cheer Pheasant	*Catreus wallichii*	IN-JK-12
Sarus Crane	*Grus antigone*	IN-JK-21 (?)
Black-necked Crane	*Grus nigricollis*	IN-JK-01, 06, 15, 18, 19
Kashmir Flycatcher	*Ficedula subrubra*	IN-JK-02, 03, 04, 08, 14

NEAR THREATENED

Oriental Darter	*Anhinga melanogaster*	IN-JK-20
Ferruginous Pochard	*Aythya nyroca*	IN-JK-05, 09, 19, 20
Long-billed Bush-warbler	*Bradypterus major*	—

that the individual wetlands are not important. All the sites are part of the IBA and should be well protected. Otto Pfister has suggested that the Changthang region could be divided into three areas, from the conservation point of view. These areas are: Pangong Tso region (northern Changthang Wilderness Area, including Pangong Tso, Chushul, Harong, and Lungparma), Hanle region (eastern Changthang Wilderness Area, including Hanle plain, Lalpari, Staklung, and Fukche) and Tso Moriri region (western Changthang Wilderness Area, including Tso Kar plain, Puga, Tso Moriri, and Chumur).

The Jammu region has mixed types of forest. The districts of Udhampur, Jammu, Kathua, Reasi, and Poonch are very important for the species of Biome 8. Detailed studies have not been conducted and very little is known about the birds here, except for a couple of rapid surveys by Baba (Wildlife Warden) in 1999–2000. The area between 500 and 1500 msl is particularly denuded, and certainly requires more study (Trevor Price, *pers. comm.* 2011). Ramnagar Wildlife Sanctuary is one of the important IBAs in Jammu region. It harbours 30 species of breeding birds, besides a variety of other important biodiversity.

Chestnut Thrush *Turdus rubrocanus* is resident in the Himalaya

Threatened Species

In Jammu & Kashmir, historically, 18 globally Threatened species have been recorded (till 2011), such as the Siberian Crane *Grus leucogeranus* from Leh, White-headed Duck *Oxyura leucocephala* and Lesser White-fronted Goose *Anser erythropus* from Wular Lake, Baikal Teal *Anas formosa* from Mirgund reservoir, Marbled Teal *Marmaronetta angustirostris* from Wular Lake and Mirgund reservoir, and Pallas's Fish Eagle *Haliaeetus leucoryphus* from Wular Lake, Haigam Rakh, Leh, Hokarsar, Chushul, Marbul pass, Tso Moriri Lake, and Hanle. The Greater Spotted Eagle *Aquila clanga* was also reported from Bhaderwah in Jammu (BirdLife International 2001) and the Eastern Imperial Eagle *Aquila heliaca* from Kashmir Valley. Cheer Pheasant (*Catreus wallichii*) was reported from Limber Wildlife Sanctuary and Kishtwar National Park. Sarus Crane *Grus antigone* was reported from Kathua district, at Kishanpur Garuna Wetland Reserve (Choudhury *et al.* 1999, Sahi 1993, Sundar 1999). Eastern Stock Pigeon *Columba eversmanni* was reported from Limber Wildlife Sanctuary (Javed 1992). Some of these species are still found in J&K, but birds such as the Siberian Crane, the White-headed Duck, the Lesser White-fronted Goose, or the Marbled and Baikal Teal have not been reported recently.

In recent years, 15 globally Threatened and Near Threatened species have been recorded from the IBAs in Jammu & Kashmir. Most of the species are widespread, such as Pallas's Fish-eagle, Eastern Imperial Eagle, White-backed Vulture and Greater Spotted

Eagle. The Black-necked Crane is found in the Changthang plateau in small numbers, their main population is in Tibet, and a wintering population of *c.* 300 individuals was seen in Bhutan (BirdLife International 2001, Rahmani 2012). Along with neighbouring Himachal Pradesh, Jammu & Kashmir is extremely important for the long-term survival of Western Tragopan *Tragopan melanocephalus* and Cheer Pheasant *Catreus wallichii*. The main breeding population of the Kashmir Flycatcher *Ficedula subrubra* is found in this state. It has been recorded breeding in Overa Wildlife Sanctuary (Price & Jamdar 1990) and in Dachigam National Park in 2003 (Khursheed Ahmad, *unpubl.*, Trevor Price, *pers. comm.* 2012).

THREATENED SPECIES FOR WHICH JAMMU & KASHMIR IS IMPORTANT

Pallas's Fish-Eagle *Haliaeetus leucoryphus* Vulnerable

This species was once very common in the valley of Kashmir. It has probably declined, although the recent dearth of records is partly the result of reduced accessibility (BirdLife International 2001). This bird has been reported from Wular Lake in the past (Unwin 1897, Loke 1946, Ludlow & Kinnear 1933–34); Haigam Rakh (Scott *et al.* 1989); Leh, Ladakh (Ludlow & Kinnear 1933–34); Hokarsar (Scott *et al.* 1989, Khursheed Ahmad, *pers. comm.* 2007), Srinagar, Chushul (Meinertzhagen 1927); Tso Moriri Lake, Ladakh (Osmaston 1925); and Hanle (Oberholser 1900). Now it is reported only from Haigam Rakh, Hokarsar, and Wular Lake and associated marshes (Khursheed, *unpubl.* 2007). In Ladakh, Pfister (2004) found it to be a rare passage migrant in spring (May-June) through the lower reaches of the valley of central Ladakh and the high altitude wetlands of eastern Ladakh upto 4,600 msl.

SACHIN RAI

Pallas's Fish-eagle *Haliaeetus leucoryphus* has declined all over India, including J&K

INTRODUCTION

Western Tragopan *Tragopan melanocephalus* Vulnerable

The Western Tragopan is classified as Vulnerable because its sparsely distributed small population is declining and becoming increasingly fragmented in the face of continuing forest loss and degradation throughout its restricted range (BirdLife International 2001). This species has been reported from Wular (Knox & Walters 1994); Lolab Valley (Lawrence 1895, Baker 1921 – 30); Limber Valley Wildlife Sanctuary (Kaul 1989, Qadri *et al.* 1990, Javed 1992, Akhtar *et al.* 1994); and Kishtwar National Park (Ward 1906 – 08). In J&K, the bird has historically been distributed only along the Pir Panjal Range (which borders the valley with Himachal) but not towards the inner Greater Himalayan Range (Suhail Intesar, *unpubl.* 2011). It has recently been recorded by Riyaz Ahmad from Badanari, Kalamund, and Hingli locations in Kalamund-Tatakuti Area and Godinuk and Pathra locations in Khara Gali Area of Poonch along southern slopes of Pir Panjal (Riyaz Ahmad, *pers. comm.* 2011).

THAKUR DALIP SINGH

Male Western Tragopan *Tragopan melanocephalus*

Cheer Pheasant *Catreus wallichii* Vulnerable

The Cheer Pheasant's small population is naturally fragmented, because it lives in small patches of successional grasslands. Human population pressure, hunting, and changing patterns of land use are leading to its decline, qualifying it as Vulnerable (BirdLife International 2001, 2011). This bird was reported from Kishtwar National Park (Ward 1906-08) and Limber Valley WLS. There is no recent record of a sighting from any area in its range in Jammu & Kashmir (Khursheed, *unpubl.* 2011).

Black-necked Crane *Grus nigricollis* Vulnerable

This species has a small, declining population as a result of the loss and degradation of wetlands, changing agricultural practices and increased human activity in its breeding and wintering grounds. These factors qualify it as Vulnerable (BirdLife International 2001). About 80 individuals are seen at present in the Ladakh region, with about 15 breeding pairs (Chandan *et al*. 2008b).

Kashmir Flycatcher
Ficedula subrubra Vulnerable

This migratory flycatcher has a small population and breeding range, which is also severely fragmented as a result of the destruction of temperate and mixed deciduous forests. It therefore qualifies as Vulnerable (BirdLife International 2001). The bird is reported from Lolab Valley and Dachigam National Park (Gauntlett 1972, Ahmad 1999), and 15 ringed, April – July 1989 (BNHS ringing data); Rampur-Rajpur Valley, above Wular Lake, "many pairs" (Bates & Lowther 1952), Overa Wildlife Sanctuary (Jamdar 1987, Price & Jamdar 1990), and Pahalgam. Though it is confirmed from four IBAs, its breeding records are available only from Overa WLS (Price & Jamdar 1990) and Dachigam National Park (Khursheed Ahmad, *unpubl*. 2003).

Hunting is the major threat to Cheer Pheasant *Catreus wallichii*

Long-billed Bush-warbler *Bradypterus major* Near Threatened

In 1994, the Long-billed Bush-warbler *Bradypterus major* was listed as Vulnerable (Collar *et al*. 1994) which was later changed to Near Threatened (BirdLife International 2001) where it has remained. It has a narrow distribution in the Western Himalaya in Pakistan and India, and in Xinjiang province of western China (Koelz 1940). In India, its main stronghold is Kashmir, with possible occurrence in Himachal Pradesh and Uttarakhand, although it has not been recorded in recent years (Rahmani 2012). It was reported from Sonamarg to Baltal in the Kashmir Valley (Bates & Lowther 1952). It has not been recorded from any IBA. Trevor Price (*pers. comm.* 2012) has not seen it in Overa.

Restricted Range species

In the Western Himalaya (Endemic Bird Area 128), the main habitats are the Temperate Coniferous or Broadleaf Forest, Subalpine Forest and Montane Grassland. These habitats have 11 restricted-range avian species between 1,500 and 3,600 msl. Of these, four are globally Threatened (Stattersfield *et al*. 1998, BirdLife International 2001). In Jammu & Kashmir 10 are found, the exception being Himalayan Quail *Ophrysia superciliosa* (thought to be extinct) which was distributed in long grass and brushwood on steep hillsides and was reported from Uttarakhand more than 100 years ago. Other restricted-range species in Jammu & Kashmir are the Western Tragopan in the dense undergrowth in Coniferous, Mixed and Oak Forests between 1,350 msl (in winter) and

Fire-fronted Serin *Serenus pusillus,* also called Gold-fronted Finch, is fairly common from 2,000 to 4,600 msl in Western Himalaya

DHRITIMAN MUKHERJEE

3,600 msl; Cheer Pheasant on steep grassy slopes, Open Coniferous or Deciduous Forests, appears to like early successional habitats between 1,400 and 3,500 msl; Brooks's Leaf-warbler *Phylloscopus subviridis* has not been reported from any of the IBAs from its habitat of Coniferous and Mixed Forest in drier, cooler areas between 2,100 and 3,600 msl; Tytler's Leaf-warbler *Phylloscopus tytleri* is reported from Dachigam National Park in Coniferous Forest, with dwarf willows and birches near the tree line between 2,400 and 3,100 msl. Breeding of the Kashmir Flycatcher *Ficedula subrubra* is restricted to Kashmir, but it winters down to the south of India in Tamil Nadu, Kerala, and in Sri Lanka. In Jammu & Kashmir, it is reported from Dachigam National Park (Gauntlett 1972, Ahmad 1999), Dehra Gali in Jammu (Tahir Shawl, *pers. comm.* 2003), Gulmarg Wildlife Sanctuary, Overa-Aru Wildlife Sanctuary, and Hirpora (Tahir Shawl, *pers. comm.* 2003) in Temperate Mixed Broadleaf Forest, especially where there is dense growth of *Parrotia* between 1,800 and 2,700 msl. White-cheeked Tit *Aegithalos leucogenys* is not reported from any IBA but Spectacled Finch *Callacanthis burtoni* is regularly seen in Overa (Price & Jamdar 1990, Price *et al.* 2003). The White-throated Tit *Aegithalos niveogularis* has been regularly seen in Overa Wildlife Sanctuary near the tree line (Price *et al.* 2003). Orange Bullfinch *Pyrrhula aurantiaca* and Kashmir Nuthatch (*Sitta cashmirensis*) are both recorded from Dachigam National Park (Khursheed Ahmad, *unpubl.* 2012). The Orange Bullfinch is regularly seen in Dachigam National Park and Overa-Aru Wildlife Sanctuary in Open Coniferous and Mixed Forest between 1,600 msl (in winter) and 2,700 to 3,300 msl during summer. Kashmir Nuthatch *Sitta cashmirensis*, though reported only from Overa-Aru (Price & Jamdar 1990) has also been recorded as sighted from a subalpine meadow at *c.* 3,300 msl in Dachigam (Ahmad 1999, Khursheed Ahmad, *unpubl.* 2003).

Biome 5

Ladakh lies in Biome 5 (Eurasian High Montane – Alpine and Tibetan) where BirdLife International (undated) has identified 48 species that represent the bird assemblages of this biome. Based on the checklist of Pfister (2004), 25 species are found here. In or near the Changthang area, thick stands of *Hippophae rhamnoides* and other vegetation in the Markha and Chang Chu Valleys provide important habitats for large numbers of wintering passerines such as Guldenstadt's Redstart *Phoenicurus erythrogaster*, Common Great Rosefinch *Carpodacus rubicilla*, Streaked Great Rosefinch *C. rubicilloides*, Black-throated Thrush *Turdus ruficollis*, Stoliczka's Tit-warbler *Leptopoecile sophiae*, Robin Accentor *Prunella rubeculoides*, and Brown Accentor *Prunella fulvescens* (Mallon 1987, 1989).

THREATS AND CONSERVATION ISSUES

As elsewhere in the country, heavy human interference and overexploitation of forests has had an adverse impact on wildlife and their habitats. Human-wildlife conflict is a huge problem these days, partly a result of intensive human interference even in protected areas. Thus the victim is wildlife in general, and threatened and endemic species in particular. Among the several Endangered species inhabiting this region is the Hangul, which is now confined to Kashmir. Musk Deer and Markhor, Cheer Pheasant, Western

Paramilitary and military forces have adopted Black-necked Crane as a conservation symbol of the Ladakh region

Sulphur-bellied Warbler *Phylloscopus griseolus* is found in Kashmir from 2,400 to 4,500 msl in summer, while it winters in central India

Tragopan, and Black-necked Crane are other Threatened species that require immediate management and conservation actions.

Wildlife conservation in general and Threatened species (particularly Hangul) conservation has been given priority in all management plans, beginning with the first one drafted in 1971 by Collin Holloway. The population of Hangul in Kashmir has been reduced from an estimated 2000 individuals in 1947 to about 140-170 individuals at

White-throated Tit *Aegithalos niveogularis* is endemic to the Himalaya from north Pakistan (Kaghan Valley) to Uttarakhand and Central Nepal from 2,400 to 4,000 msl. In winter it can be seen down to 1,800 msl

INTRODUCTION

present (Ahmad *et al.* 2009). Livestock grazing is a major problem in all the protected areas and IBAs in J&K. Even in the prestigious Dachigam NP, a sheep farm is present. Despite repeated attempts, the Wildlife Department has not been successful in relocating this farm outside the national park. Other problems include the lack of coordination between the many different departments that hold stakes in the Park (Animal Husbandry, Hospitality and Protocol, PWD, Irrigation and Water Works, Electricity, Telephones, Agriculture and Fisheries). Disturbance to wildlife is also caused by visitors driving noisily along the 5 km stretch of road to the VIP lodge at Draphama (Gruisen 1983). At present, the army and paramilitary forces have a base inside the National Park, and they

WWF-India, along with MoEF and BNHS organized a very successful regional workshop on the conservation of Black-necked Crane. L-R Mr. Ravi Singh, CEO of WWF-India,
Mr. Dasho Paljor Dorji, Mr. Jairam Ramesh, then Minister of Environment and Forests, and
Mr. Jagdish Kishwan, MoEF

not only occupy the accommodation meant for frontline wildlife staff but also cause disturbance to the Hangul habitats, particularly during its breeding season. Similar is the case with other protected areas and IBAs.

In Gulmarg, the tourist industry depends on the surrounding forest for fuelwood. The forest also suffers, like most sanctuaries of Jammu & Kashmir, from the invasion of nomadic graziers in certain months.

Overa, Lardi, and Dahwattoo villages, with a total human population of nearly 4,500 are situated close to the southern boundary of Overa-Aru WLS. Constant vigil is required to prevent encroachment into the sanctuary (Suhail 2000). The upper parts of this IBA suffer massive grazing pressure during the summer from local and nomadic graziers. Firewood collection is a major problem due to the increase in human population. Many villagers have to stock large amounts of dry wood to see them through winter, though increasingly they are using natural gas. During the dry summer months, forest fire is also an issue. Many times fires are deliberately set by graziers to remove dry, unpalatable, or inedible coarse grasses.

ASAD R. RAHMANI

In Ladakh, unplanned developmental activities are the main concern in the Changthang region. Alteration of marshes by willow plantation and construction of roads is directly affecting the breeding grounds of the Vulnerable Black-necked Crane. Increasing settlements near the crane's nesting habitats have resulted in an increase in the feral dog population, a major predator on crane eggs in Changthang.

Overall, the key threats to birds and other biodiversity of the state are habitat encroachment, overgrazing by livestock, tourism, firewood collection, and forest fires.

Involving local people is important for long-term conservation initiatives

INTESAR SUHAIL

Tulip fields now attract a large number of tourists to the Kashmir Valley. Bird tourism has still not picked up although it has great potential as J&K has some of the most valuable Wildlife Sanctuaries and IBAs in India

Overgrazing by livestock has become a major problem for ground-nesting birds

REFERENCES

Ahmad, K. (1999) Birds of Dachigam National Park. *Newsletter for Birdwatchers* 39(2): 22-24.

Ahmad, K., Sathyakumar, S. and Qureshi, Q. (2009) Conservation status of the last surviving population of Hangul (*Cervus elaphus hanglu*) in Kashmir. *JBNHS* 106(3): 245-255.

Akhtar, A., Prakash, V. and Javed, S. (1994) The Western Tragopan: bird of the Himalaya. *Sanctuary (Asia)* 14(2): 44-49.

Baker, E.C.S. (1921-30) *The Game-birds of India, Burma and Ceylon*. Bombay Natural History Society, Bombay.

Bates, R.S.P. and Lowther, E.H.N. (1952) *Breeding Birds of Kashmir*. Oxford University Press, Oxford, UK.

BirdLife International (2001) *Threatened Birds of Asia: The BirdLife International Red Data Book*. 2 vols. BirdLife International, Cambridge, UK.

BirdLife International (undated) *Important Bird Areas (IBAs) in Asia: Project briefing book*. BirdLife International, Cambridge, UK. Unpublished.

BirdLife International (2011) IUCN Red List for Birds. Downloaded from http://www.birdlife.org on 25/07/2011.

Chandan, P., Gautam, P. and Chatterjee, A. (2006) Nesting sites and breeding success of Black-necked Crane *Grus nigricollis* in Ladakh, India. Pp. 311-314. In: Boere, G.C., Galbraith, C.A. and Stroud, D.A. (eds) *Waterbirds Around the World*. The Stationery Office, Edinburgh, UK.

Chandan, P., Chatterjee, A. and Gautam, P. (2008a) Management Planning of Himalayan High Altitude Wetlands. A case study of Tsomoriri and Tsokar Wetlands in Ladakh, India. *Proceedings of Taal 2007: The 12th World Lake Conference*, 1446-1452.

Chandan, P., Abbas, M. and Gautam, P. (2008b) *Field Guide: Birds of Ladakh*. WWF-India and Department of Wildlife Protection, Government of J&K, Srinagar.

Choudhury, B.C., Kaur, J. and Gopi Sundar, K.S. (1999) Sarus Crane Count-1999. Wildlife Institute of India, Dehradun.

Collar, N.J., Crosby, M.J. and Stattersfield, A.J. (1994) *Birds to Watch 2: the world checklist of threatened birds*. BirdLife International Cambridge, UK.

Gauntlett, F.M. (1972) Notes on some Kashmir birds. *JBNHS* 69: 591-615.

Gujja, B., Chatterjee, A., Gautam, P. and Chandan, P. (2003) Wetlands and Lakes at the Top of the World. *Mountain Research and Development* 23(3): 219-221.

Gruisen, J. van (1983) The Hangul, Dachigam's endangered deer. *Sanctuary (Asia)* 3: 114-131.

Islam, M.Z. and Rahmani, A.R. (2004) *Important Bird Areas in India: Priority sites for conservation*. Indian Bird Conservation Network, BNHS, Mumbai, India and BirdLife International, UK. Oxford University Press, Mumbai. Pp. xvii + 1133.

Islam, M.Z. and Rahmani, A.R. (2008) *Existing and Potential Ramsar Sites in India*. Indian Bird Conservation Network, BNHS, BirdLife International and Royal Society for the Protection of Birds. Oxford University Press. Pp. 592.

Jamdar, N. (1987) Additions to the birds of Point Calimere, south India. *JBNHS* 84: 206.

Javed, S. (1992) Birds of Limber Valley forest (Jammu and Kashmir). *Newsletter for Birdwatchers* 32(5/6): 13-15.

Kaul, R. (1989) Western Tragopan surveys in the Limber Valley, Kashmir, India. *WPA News* 26: 12-14.

Knox, A.G. and Walters, M.P. (1994) *Extinct and endangered birds in the collections of the Natural History Museum*. British Ornithologists' Club, London, UK.

Koelz, W. (1940) Notes on the birds of Zanskar and Purig, with appendices giving new records from Ladakh, Rupshu and Kulu. *Pap. Michigan Acad. Sci. Arts Letters* 25: 297-322.

Lawrence, W.R. (1895) *The Valley of Kashmir*. H. Frowde, Oxford University Press, London.

Loke, W.T. (1946) A bird photographer in Kashmir. *JBNHS* 46: 431-436.

Ludlow, F. and Kinnear, N.P. (1933-34) A contribution to the ornithology of Chinese Turkestan. *Ibis* 13(3): 240-259, 440-473, 658-694; 13(4): 95-125.

Mallon, D.P. (1987) The winter birds of Ladakh. *Forktail* 3: 27-41.

Mallon, D.P. (1989) An ecological survey of the protected area network in Ladakh. Report to the Department of Wildlife Protection, Jammu and Kashmir. Unpublished.

Mathew, K.M. (ed) (2003) Manorama Yearbook 2003. Malayale Manorama, Kottayam.

Meinertzhagen, R. (1927) Systematic results of birds collected at high altitudes in Ladak and Sikkim. *Ibis* (12)3: 363-422, 571-633.

Ministry of Environment and Forests (2001) *State of Forest*. Forest Survey of India, Dehradun.

Oberholser, H.C. (1900) Notes on birds collected by Dr W.L. Abbott in Central Asia. *Proc. U.S. Natl. Mus.* 22: 205-228.

Osmaston, B.B. (1925) The birds of Ladakh. *Ibis* (12)1: 663-719.

Pfister, O. (1998) *The breeding ecology and conservation of the Black-necked Crane* (Grus nigricollis) *in Ladakh, India*. Unpublished Thesis. University of Hull, Hull, UK. Pp. 136.

Pfister, O. (2004) *Birds and Mammals of Ladakh*. Oxford University Press, New Delhi.

Price, T.D. and Jamdar, N. (1990) The breeding birds of Overa Wildlife Sanctuary, Kashmir. *JBNHS* 87: 1-15.

Price, T., Zee, J., Jamdar, K. and Jamdar, N. (2003) Bird species diversity along the Himalaya: A comparison of Himachal Pradesh with Kashmir. *JBNHS* 100: 394-410.

Qadri, M.Y., Kaul, R. and Iqbal, M. (1990) Status of pheasants in Kashmir with special reference to endangered species. Pp. 124-128. In: Hill, D.A., Garson, P.J. and Jenkins, D. (eds) *Pheasants in Asia 1989*. World Pheasant Association, Reading, UK.

Rahmani, A.R. (2012) *Threatened Birds of India: Their Conservation Requirements*. Indian Bird Conservation Network, Bombay Natural History Society, Royal Society for the Protection of Birds and BirdLife International. Oxford University Press, Mumbai. Pp. 864.

Sahi, D.N. (1993) Wildlife Conservation sites in Kashmir Himalayas. *Tigerpaper* 20(2): 28-31.

Scott, D.A. (1989) *A Directory of Asian Wetlands*. IUCN, Gland, Switzerland, and Cambridge, UK.

Stattersfield, A.J., Crosby, M.J., Long, A.J. and Wege, D.C. (1998) *Endemic Bird Areas of the World: Priorities for Biodiversity Conservation*. BirdLife Conservation Series No. 7. BirdLife International, Cambridge, UK.

Suhail, I. (2000) Overa-Aru Wildlife Sanctuary: Management Plan: 2001-2006. Department of Wildlife Protection, Srinagar.

Sundar, K.S.G. (1999) The Sarus in Jammu, the Fulvous Whistling-duck in north Bengal and birds in Pondicherry University Campus - a reply. *Newsletter for Birdwatchers* 39(3): 41-43.

Unwin, W.A. (1897) Late stay of wildfowl. *JBNHS* 11: 169.

Ward, A.E. (1906-08) Birds of the provinces of Kashmir and Jammu and adjacent districts. *JBNHS* 17: 108-113, 479-485, 723-729, 943-949; 18: 461-464.

Wetland International (2006) *Waterbirds Population Estimates: Fourth Edition*. Wetlands International Global Series No. 12, Wageningen, The Netherlands.

DESCRIPTION OF THE IMPORTANT BIRD AREAS

CHUSHUL MARSHES

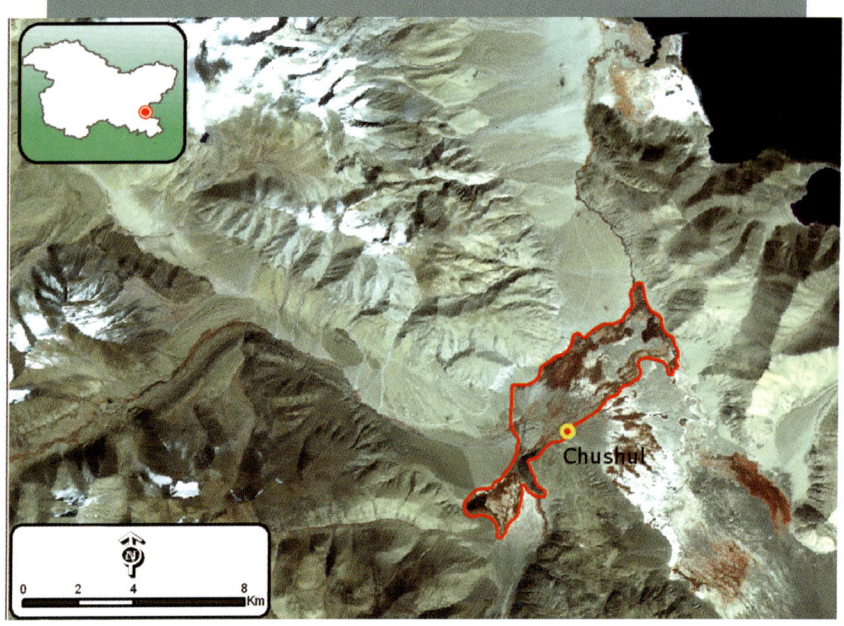

IBA Site Code	IN-JK-01
District	Leh, Ladakh
Coordinates	33⁰ 33' 07.6" N, 78⁰ 38' 58" E to 35⁰ 35' 03" N, 78⁰ 45' 00" E
Ownership	State Wildlife Department
Area	1,500 ha
Altitude	4,329 msl
Precipitation	100 mm + snowfall
Temperature	-30˚C to 30˚C
Biogeographic Zone	Trans-Himalaya
Habitats	Aquatic: shallow marshes, flooded valley, grassland, plains, wet meadows

IBA CRITERIA	: A1 (Threatened Species)
PROTECTION STATUS	: Wetland Reserve.

GENERAL DESCRIPTION

The Chushul marshes lie within Changthang, east and west of Chushul village near the Chinese border. The habitats consist of shallow ponds, marshes, borax plains, and wet meadows in a broad sandy valley. Springs and streams flowing down into the valley from the Ladakh range create these ponds, pools, and marshes. Some streams terminate on the sandy plains in stagnant pools which become more saline as they evaporate. Others carry enough water to flow into the Pangong Tso. Most of the ponds and marshes remain frozen from December to March. The principal vegetation consists of species of *Hydrilla* and *Myriophyllum* in the ponds, and *Carex*, other sedges, and grasses in the marsh. The surrounding steppe is dominated by *Caragana*.

Chushul contains shallow ponds, marshes, borax plains, and wet meadows, thus creating a mosaic of habitats for a large number of species

AVIFAUNA

The wetland is an important breeding area for several species of waterfowl, such as Ruddy Shelduck *Tadorna ferruginea*, Lesser Sand Plover *Charadrius mongolus*, Common Redshank *Tringa totanus*, and Common Tern *Sterna hirundo*. Three to four pairs of Black-necked Crane *Grus nigricollis* also inhabit the Chushul marshes (Hussain & Pandav 2001, Rauf Zargar, *pers. comm.* 2003, Chandan *et al.* 2006). Some pairs of Bar-headed Geese *Anser indicus* also breed here. During our survey in July 2007, four pairs of Black-necked Crane were sighted in Chushul marshes and adjoining Tsigul-so and Lung Parma marshes, including two pairs of Black-necked Crane with two chicks each sighted at Tsigul-so and Lung Parma marshes foraging alongside livestock (Khursheed Ahmad, *unpubl.* 2007). Besides 24 pairs of Bar-headed Geese with 49 chicks, a Golden Eagle *Aquila chrysaetos* was also recorded breeding here. Among other waterbirds, Northern Pintail *Anas acuta*, Redshank *Tringa totanus*, Common Merganser *Mergus merganser*, Ruddy Shelduck *Tadorna ferruginea*, Black Redstart *Phoenicurus ochruros*, and Common Tern *Sterna hirundo* were observed (Khursheed Ahmad, *unpubl.* 2007). Tibetan Sandgrouse *Syrrhaptes tibetanus*, and Tibetan Partridge *Perdix hodgsoniae* representing Biome 5 occur on the surrounding dry plains.

Species of concern is the Vulnerable Black-necked Crane *Grus nigricollis*.

OTHER KEY FAUNA

The other important fauna include Tibetan Wild Ass *Equus kiang*, Tibetan Argali *Ovis ammon*, Blue Sheep *Pseudois nayaur*, Tibetan Gazelle *Procapra picticaudata*, Tibetan Woolly Hare *Lepus oiostolus*, and Long-tailed Marmot *Marmota caudata* (Rauf Zargar, *pers. comm.* 2003, Khursheed Ahmad, *unpubl.* 2007).

Golden Eagle *Aquila chrysaetos* breeds in the mountains surrounding Chushul

LAND USE

○ Urban settlements
○ Nature conservation and research
○ Agriculture
○ Paramilitary (ITBP) establishments

THREATS AND CONSERVATION ISSUES

○ Plantation in wetland
○ Livestock grazing
○ Feral dogs
○ Soil erosion

The wetland lies within the proposed High Altitude Cold Desert National Park in east Ladakh. Human interference in the area, including increasing livestock grazing in and around the wetlands, poses a threat to the vegetation and causes soil erosion. Permanent human settlements cause contamination of the water, brooks are diverted for agriculture which drains the marshes, and increased garbage production attracts the Raven *Corvus corax* which, together with semi-feral dogs (locally called *yanki*s), prey on small mammals, and eggs and nestlings of waterfowl.

Freshwater marshes are rare in Ladakh, so they are a focal point for human beings as well as wildlife. *Caragana* bushes are collected by the local people to feed livestock. During religious festivals, people congregate in these wetlands and there is increased diversion of water channels for domestic use (Rauf Zargar, *pers. comm.* 2003).

KEY CONTRIBUTORS

Rauf Zargar, Bivash Pandav, S.A. Hussain, and Khursheed Ahmad.

DACHIGAM NATIONAL PARK

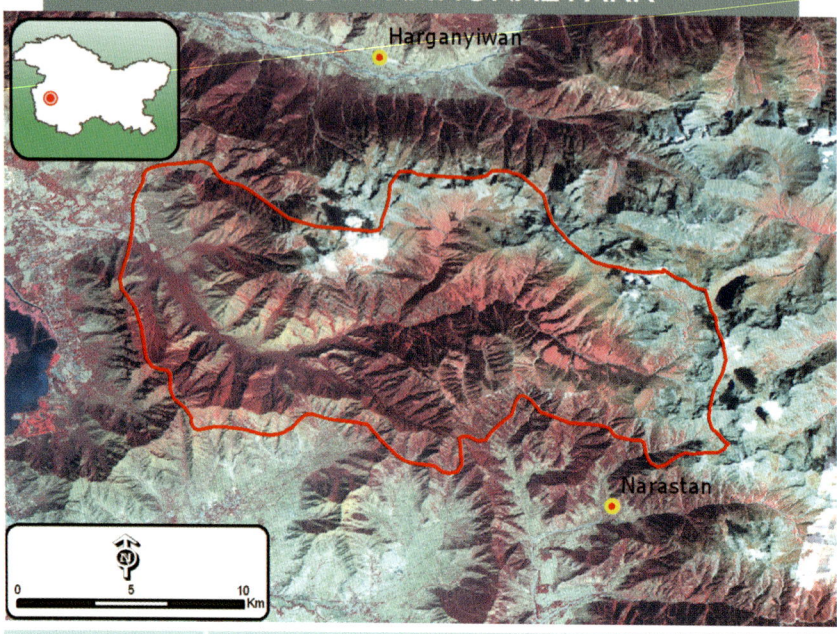

IBA Site Code	IN-JK-02
District	Srinagar, Pulwama
Coordinates	34° 12' 10" N, 74° 51' 36" E
Ownership	State Wildlife Department
Area	141 sq. km
Altitude	1,642–4,700 msl
Precipitation	32–546 mm + snowfall
Temperature	-2 °C to 32 °C
Biogeographic Zone	Northwest Himalaya 2A
Habitats	Himalayan Moist Temperate Mixed Deciduous Forest, Himalayan Moist Subalpine Mixed Coniferous Forest, Himalayan Alpine Moist Scrub, meadows

IBA CRITERIA	:	A1 (Threatened Species), A2 (Endemic Bird Area 128: Western Himalaya), A3 (Biome 5: Eurasian High Montane, Biome 7: Sino-Himalayan Temperate Forest)
PROTECTION STATUS	:	National Park, established since 1981.

GENERAL DESCRIPTION

Dachigam National Park, about 20 km from Srinagar, was established in 1910 as a hunting reserve by the Maharaja of Kashmir. The importance of Dachigam as a catchment area of Dal Lake was also recognized by the erstwhile Maharaja, who resettled 10 villages to protect the forest cover, hence the name Dachigam (*dachi* = ten, *gam* = village). After

Owing to its proximity to Srinagar town, Dachigam NP is very popular with tourists and government officials

the merger of Jammu & Kashmir with India in 1948, the management of the Park was handed over to the erstwhile Game and Fisheries Department and subsequently to the Wildlife Protection Department, which separated from the Forest Department in 1980.

Dachigam National Park lies between 34° 05' 00" N and 34° 10' 32" N, and 74° 53' 50" E and 75° 09' 16" E (Ahmad *et al.* 2009) in the Zanskar mountain range of Northwest Himalayan biogeographic zone (2A).

It is roughly rectangular, *c.* 22.5 km long, 8 km wide, and covers half the catchment area of Dal Lake (Holloway & Wani 1971). Dachigam National Park is a part of Zanskar Range which forms the northwest branch of the Central Himalayan Axis, bifurcating near Kullu (Himachal Pradesh) and terminating in the high twin peaks of Nun Kun (7,135 msl) (Holloway & Wani 1971). The fold of this mountain range has narrow gullies, and broader outer gullies locally called *nar*. Two steep ridges, one rising near Harwan Reservoir and the other to the east of New Theed, form the natural boundaries of the Park (Holloway & Wani 1971, Kurt 1978b). Dachigam nurtures a variety of vegetation because of varied habitats, density of dominant species, controlled by diverse microclimatic conditions due to the changing aspects of the undulating terrain, exposure, altitude, and above all, biotic interference.

This Park is a major catchment area of the Dal Lake, and a source of drinking water for Srinagar. The Dagwan stream which brings potable water originates from Marsar Lake in Upper Dachigam and is fed by numerous streams (Kurt 1978b) till it drains into Harwan Reservoir.

Dachigam National Park has great ecological, aesthetic, and socio-economic significance. It harbours the last surviving population of the Critically Endangered, endemic Kashmir Red Deer or Hangul *Cervus elaphus hanglu*. The Hangul, once distributed widely in the mountains of Kashmir, has reduced drastically due to habitat degradation, poaching, and indiscriminate biotic pressure. At present the population numbers 140 to 170 individuals restricted to Dachigam and the adjoining protected areas (Ahmad *et al*. 2009).

Owing to protection for over 90 years, the vegetation of Dachigam NP is in strong contrast with that outside. Despite the pressure of graziers in the Park, the vegetation is

ENDANGERED	
Egyptian Vulture	*Neophron percnopterus*

VULNERABLE	
Eastern Imperial Eagle	*Aquila heliaca*
Kashmir Flycatcher	*Ficedula subrubra*

Endemic Bird Area 128: Western Himalaya	
Tytler's Leaf-warbler	*Phylloscopus tytleri*
Kashmir Flycatcher	*Ficedula subrubra*
Orange Bullfinch	*Pyrrhula aurantiaca*
Kashmir Nuthatch	*Sitta cashmirensis*

Biome 5: Eurasian High Montane (Alpine and Tibetan)	
Himalayan Griffon	*Gyps himalayensis*
Grandala	*Grandala coelicolor*
Tickell's Leaf-warbler	*Phylloscopus affinis*
Red-mantled Rosefinch	*Carpodacus rhodochlamys*
Red-fronted Rosefinch	*Carpodacus puniceus*
Himalayan Rubythroat	*Luscinia pectoralis*

Biome 7: Sino-Himalayan Temperate Forest	
Koklass Pheasant	*Pucrasia macrolopha*
Himalayan Monal	*Lophophorus impejanus*
Streaked Laughingthrush	*Garrulax lineatus*
Variegated Laughingthrush	*Garrulax variegatus*
Long-billed Bush-warbler	*Bradypterus major*
Western Crowned Leaf-warbler	*Phylloscopus occipitalis*
Rusty-tailed Flycatcher	*Muscicapa ruficauda*
Fire-capped Tit	*Cephalopyrus flammiceps*
Simla Crested Tit	*Parus rufonuchalis*
Green-backed Tit	*Parus monticolus*
Bar-tailed Treecreeper	*Certhia himalayana*
Yellow-breasted Greenfinch	*Carduelis spinoides*
Yellow-billed Blue Magpie	*Urocissa flavirostris*

Dachigam National Park is the sole remaining habitat of Kashmir Stag or Hangul

more or less intact, although invasive species are now seen in some areas. The slopes of Dagwan Valley and the catchment areas of various nullahs in Lower Dachigam sustain almost pristine vegetation. Six major types of vegetation have been recognised. The lower areas, from 1,700 to 3,000 msl, have Broadleaf Mesophyll forests of *Acer caesium*, *Morus alba*, *Ulmus* spp., *Rhus succedanea*, *Juglans regia*, *Parrotiopsis jacquemontiana*, and conifers such as Blue Pine *Pinus wallichiana*, Spruce *Picea smithiana*, and Fir *Abies pindrow* growing in altitudinal succession (Holloway 1970). The upper reaches, from 3,000 to *c.* 4,500 msl, comprise a vegetation gradient of Subalpine forest community followed by scrub Silver Birch *Betula utilis* and Rhododendron *Rhododendron* spp., interspersed with herb-rich grasslands and meadows above 3,300 msl. This gradually merges into the zone of permanent snow, above 4500 msl (Holloway 1970, Ahmad *et al.* 2009). Habitat gradation is similar to Overa-Aru, but importantly, Dachigam starts at a lower altitude, and is an important location in the Kashmir Valley for several species not found in Overa, including generally common species such as the Yellow-billed Blue Magpie *Urocissa flavirostris* and the Grey-hooded Warbler *Phylloscopus xanthoschistos*, and Grey-winged Blackbird *Turdus boulboul* (Trevor Price, *pers. comm.* 2012).

AVIFAUNA

Dachigam NP is rich in high altitude birds. Before insurgency started in 1989, it was very popular with birdwatchers and researchers. Katti (1989) recorded 145 bird species, while Hussain (1989) noted 107 species during the BNHS Bird Migration Project. Many birds were ringed, and their identity confirmed. Ahmad (1999) recorded 45 bird species in Lower Dachigam, while Mohammad Raashid (*unpubl.* 2012) has recorded 111 species in the Lower Dachigam area.

This site is important for the globally Vulnerable Kashmir Flycatcher *Ficedula subrubra*, which was observed in Dachigam's Mixed Woodland habitats, dominated by *Rubinia pseudoacacia* plants (Ahmad 1999). Courtship behaviour was observed by Khursheed

Ahmad in 2003 in the mixed woodland habitat near the Sheep Breeding Farm (Khursheed Ahmad, *unpubl.*). This migratory flycatcher has a small population and breeding range, which is severely fragmented due to the destruction of Temperate Mixed Deciduous Forests (BirdLife International 2001). It has recently been found wintering in moderate numbers in Mukurthi NP (an IBA) in Tamil Nadu (Zarri & Rahmani 2004).

Dachigam lies in Endemic Bird Area 128 (Western Himalaya) where Stattersfield *et al.* (1998) have listed 11 restricted-range species, of which three are found here.

Dachigam represents two biomes: Biome 5 Eurasian High Montane (Alpine and Tibetan), above *c.* 3,600 msl, and Biome 7 Sino-Himalayan Temperate Forest, between 1,800 and 3,600 msl. According to the BirdLife International (undated) list of biome species, out of 48 Biome 5 species, 7 are found here, and out of 112 Biome 7 species, 13 are found.

Khursheed Ahmad recorded the migratory Northern Pintail *Anas acuta* and Mallard *Anas platyrhynchos* on June 20, 1998 at Harwan Reservoir inside the Park (Ahmad 1999). While there are many records of Mallard breeding in Kashmir (Bates & Lowther 1952), the presence of Northern Pintail in summer, so far from its known breeding range, is interesting.

Of the pheasants, Himalayan Monal *Lophophorus impejanus* and Koklass *Pucrasia macrolopha* are present. Himalayan Snowcock *Tetraogallus himalayensis* has been reported (Rodgers & Panwar 1988). Other breeding species include Bullfinch *Pyrrhula aurantiaca* and Tytler's Leaf-warbler *Phylloscopus tytleri*. The White-backed Vulture *Gyps bengalensis* reported from this IBA during IBA workshop in 2004 hence included in Islam & Rahmani (2004) has not been recorded by most birdwatchers. However, Himalayan Griffon *Gyps himalayensis*, Griffon Vulture *Neophron percnopterus*, Bearded Vulture or Lammergeier *Gypaetus barbatus* and Himalayan Golden Eagle *Aquila chrysaetos* are easily seen. The Vulnerable Eastern Imperial Eagle *Aquila heliaca* is seen during the migratory season. Kashmir Nuthatch *Sitta cashmirensis*, Large-spotted Nutcracker *Nucifraga multipunctata*, Rusty-tailed Flycatcher *Muscicapa ruficauda*, Strong-footed (Brown-flanked) Bush-warbler *Cettia fortipes*, Western Crowned Leaf-warbler *Phylloscopus occipitalis*, Variegated Laughingthrush *Garrulax variegatus*, Fire-capped Tit *Cephalopyrus flammiceps*, and Simla Crested Tit *Parus rufonuchalis* are some of the other important birds seen here. Black-and-Yellow Grosbeak *Mycerobas icterioides* and Chestnut Thrush *Turdus rubrocanus* are also seen during the winters till early spring (Ahmad 1999).

OTHER KEY FAUNA

Apart from the Hangul, Dachigam National Park is home to 15 species of mammals (Anon. 1985, Ahmad *et al.* 2009).

Himalayan Black Bear *Ursus thibetanus* is widely distributed (Kurt 1979) but the Brown Bear *Ursus arctos* is uncommon, found only in Upper Dachigam (Kurt 1979, Gruisen 1983). Dachigam harbours the maximum density of Himalayan Black Bear *Ursus thibetanus* in India (Sathyakumar *et al.* 2001). Himalayan Musk Deer *Moschus chrysogaster*, and Serow *Nemorhaedus sumatraensis* are other uncommon ungulates of the higher reaches of Dachigam. There is no recent record of Snow Leopard *Uncia uncia*, although Holloway (1970) reported seeing one. Leopard *Panthera pardus*, the major natural predator, is

Orange Bullfinch *Pyrrhula aurantiaca* is endemic to Western Himalaya

quite common. Himalayan Yellow-throated Marten *Martes flavigula*, Beech or Stone Marten *Martes foina*, Himalayan Weasel *Mustela sibirica*, Jungle Cat *Felis chaus*, Golden Jackal *Canis aureus*, and Red Fox *Vulpes vulpes* are some of the smaller predators. Long-tailed Marmot *Marmota caudata* and Himalayan Mouse Hare *Ochotona roylei* and birds are the main prey. Mir Mansoor (*pers. comm.*) recently reported and photographed the Himalayan Wolf *Canis lupus chanco* from the Park.

Of the seven Langur species described by Groves (2001), Nepal Langur *Semnopithecus schistaceus* is found. It moves around in troupes, often of 60 or more (Gruisen 1983). Small Indian Civet *Viverricula indica* was recently recorded in the Park. Himalayan Pit Viper *Gloydius himalayanus* is the only poisonous snake commonly seen.

LAND USE
- ○ Nature conservation and research
- ○ Ecotourism and nature education
- ○ Watershed conservation
- ○ Various government organisations (Sheep Breeding Farm, Fisheries Farm), Hospitality & Protocol VVIP Guest House, PHE, Irrigation, Power, Telephone and PWD (R&B) department huts and staff, above all paramilitary establishments and their dogs

THREATS AND CONSERVATION ISSUES
- ○ Overgrazing
- ○ Unregulated tourism
- ○ Lack of coordination among different departments involved in the Park
- ○ Poaching
- ○ Disturbance due to movement of staff of various departments and paramilitary personnel inside the Park
- ○ Dogs of paramilitary units

An estimated 10,000 sheep and 5,000 water buffalo belonging to Chopans, Gujjars, Bakarwals, and Banyaris used to graze on the alpine pastures in summer, and wood and grass was collected by local villagers (Kurt 1978a,b, 1979). There are no longer any permanent settlements within the Park. Nomadic Gujjars and Bakarwals, however, continue to occupy the Dagwan pastures of Upper Dachigam in summer (Ahmad *et al.* 2009) and grazing is the main problem. The Government owned Sheep Breeding Farm is also a problem. Though the decision to relocate it outside the park was taken long ago, it has not been implemented. The presence of army and paramilitary forces causes much disturbance by their movements, always in convoys with escort vehicles.

KEY CONTRIBUTORS
Rashid Y. Naqash, M.S. Bacha, Intesar Suhail, Khursheed Ahmad, and Trevor Price.

DEHRA GALI (DKG) FOREST

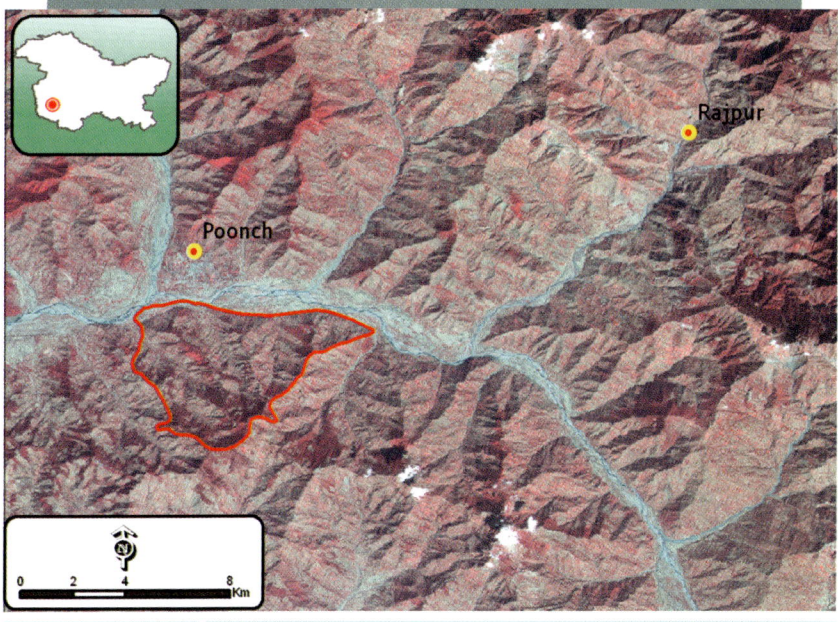

IBA Site Code	IN-JK-03
District	Poonch, Rajouri
Coordinates	33° 34' 00" N, 74° 24' 00" E
Area	1,800 ha
Altitude	1,650–2,396 msl
Precipitation	1150 mm + snowfall
Temperature	-4 °C to 25 °C
Biogeographic Zone	Himalaya
Habitats	Himalayan Wet Temperate, Subtropical Pine Forest, Subtropical Broadleaf Forest
IBA CRITERIA	: A1 (Threatened Species), A2 (Endemic Bird Area 128: Western Himalaya)
PROTECTION STATUS	: Not officially protected.

GENERAL DESCRIPTION

This IBA site in the foothills of the Pir Panjal Range is located about 32 km northeast of Rajouri town and 190 km from Jammu city. It lies in the temperate region and experiences heavy snowfall in winter, while summer is moderate. Realising the importance of this site as a key habitat for birds, the Jammu & Kashmir Forest Department is promoting it as a major ecotourism destination.

VULNERABLE	
Western Tragopan	*Tragopan melanocephalus*
Kashmir Flycatcher	*Ficedula subrubra*

Endemic Bird Area 128: Western Himalaya	
Western Tragopan	*Tragopan melanocephalus*
Kashmir Flycatcher	*Ficedula subrubra*

AVIFAUNA

Not much information on birdlife is available, but as the habitat is still suitable, it could be an excellent IBA (Shawl 1997). Intesar Suhail (*pers. comm.* 2003) has seen the Kashmir Flycatcher *Ficedula subrubra*, a globally Vulnerable species according to BirdLife International (2001). Another species of special conservation concern is the Western Tragopan *Tragopan melanocephalus*. The Himalayan Monal *Lophophorus impejanus* is also found here, perhaps in considerable numbers.

Some other key species of birds found in DKG include Lammergeier *Gypaetus barbatus*, Golden Eagle *Aquila chrysaetos*, Himalayan Griffon *Gyps himalayensis*, Common Kestrel *Falco tinnunculus*, Black-headed Jay *Garrulus lanceolatus*, Oriental Turtle-dove *Streptopelia orientalis*, and Scaly-bellied Woodpecker *Picus squamatus*.

OTHER KEY FAUNA

The area exhibits a rich and diverse faunal composition: Himalayan Black Bear *Ursus thibetanus*, Brown Bear *Ursus arctos*, Himalayan Musk Deer *Moschus chrysogaster*,

INTESAR SUHAIL

Dehra Gali located in Pir Panjal has diverse habitat types due to its great altitude range

WesternTragopan is one of the threatened bird species found in Dehra Gali

Leopard *Panthera pardus*, Himalayan Yellow-throated Marten *Martes flavigula*, Beech or Stone Marten *Martes foina*, Himalayan Weasel *Mustela sibirica*, Jungle Cat *Felis chaus*, Golden Jackal *Canis aureus*, Red Fox *Vulpes vulpes*, Long-tailed Marmot *Marmota caudata*, and Himalayan Mouse Hare *Ochotona roylei*.

LAND USE
- ○ Agriculture
- ○ Tourism
- ○ Grazing

THREATS AND CONSERVATION ISSUES
- ○ Illegal felling
- ○ Grazing
- ○ Poaching
- ○ Encroachment

One the major threats to the biodiversity of the area which has emerged in recent times is increasing vehicular traffic on the road passing through the forest. This is because of the road being a link to the historic Mughal Road connecting Jammu with Kashmir Valley.

The other major threats to this area are unplanned and unregulated developmental activities, overgrazing, illicit cutting of trees, poaching, and encroachment.

KEY CONTRIBUTORS
Tahir Shawl, Pankaj Chandan, and Intesar Suhail.

GULMARG WILDLIFE SANCTUARY

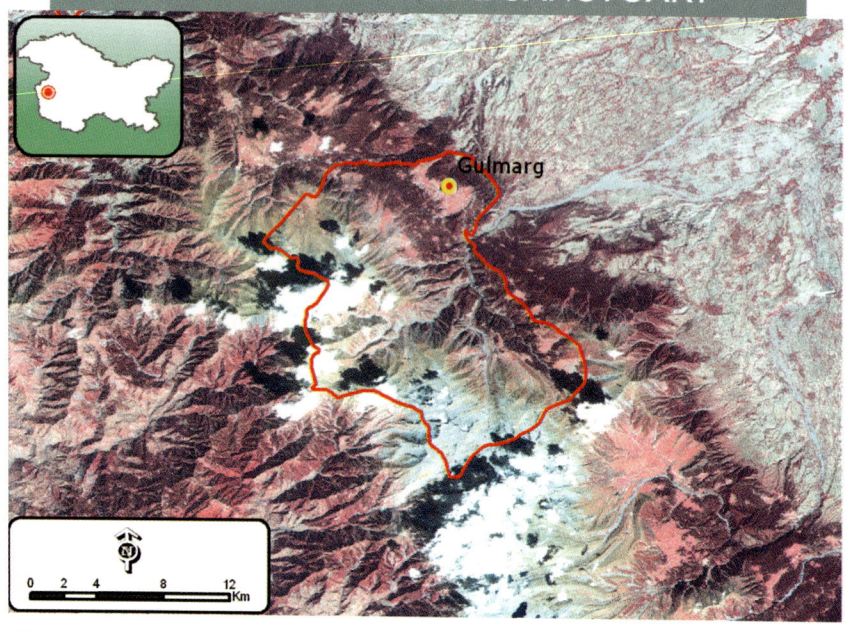

IBA Site Code	IN-JK-04
District	Baramulla
Coordinates	34° 15' 48" N, 74° 13' 23" E
Ownership	State
Area	13,925 ha
Altitude	2,400–4,300 msl
Annual Snowfall	> 14 m
Temperature	-8° C to 37 °C
Biogeographic Zone	Himalaya
Habitats	Himalayan Wet Temperate, Alpine Moist Scrub, Alpine Moist Pasture

IBA CRITERIA	: A1 (Threatened Species), A2 (Endemic Bird Area 128: Western Himalaya), A3 (Biome 5: Eurasian High Montane, Biome 7: Sino-Himalayan Temperate Forest)
PROTECTION STATUS	: Wildlife Sanctuary, established in March 1987.

GENERAL DESCRIPTION

As the name indicates, the Sanctuary surrounds the famous tourist resort of Gulmarg. It was proposed to be a biosphere reserve in 1981, but that did not work out, but the Department of Wildlife Protection declared it a Sanctuary in 1987. The Sanctuary lies on the northeast side of the Pir Panjal Range, c. 50 km southwest of Srinagar. It encompasses the upper catchment area of Ferozpur Nullah and the forests that surround

Biome 5: Eurasian High Montane	
Himalayan Griffon	*Gyps himalayensis*
Himalayan Snowcock	*Tetraogallus himalayensis*
Ibisbill	*Ibidorhyncha struthersii*
Tickell's Leaf-warbler	*Phylloscopus affinis*

Biome 7: Sino-Himalayan Temperate Forest	
Himalayan Monal	*Lophophorus impejanus*
Koklass Pheasant	*Pucrasia macrolopha*
Himalayan Pied Woodpecker	*Dendrocopos himalayensis*
Himalayan Rubythroat	*Luscinia pectoralis*
Indian Blue Robin	*Luscinia brunnea*
Streaked Laughingthrush	*Garrulax lineatus*
Variegated Laughingthrush	*Garrulax variegatus*
Large-billed Leaf-warbler	*Phylloscopus magnirostris*
Western Crowned-warbler	*Phylloscopus occipitalis*
Simla Crested Tit	*Parus rufonuchalis*
Spot-winged Crested Tit	*Parus melanolophus*
Green-backed Tit	*Parus monticola*
White-cheeked Nuthatch	*Sitta leucopsis*
Bar-tailed Tree-creeper	*Certhia himalayensis*

the Gulmarg meadows. It is bounded to the north by Jhelum Valley Forest Division, to the south and west by Poonch and Pir Panjal Forest Divisions, and to the east by Drang village. The terrain is steep, becoming precipitous in the upper reaches of Ferozpur Nullah. The underlying rocks are predominantly Panjal volcanics, with well exposed acidic lava flows. Shale, limestone, slate, and quartzite occur throughout the tract (Anon. 1987, Bacha 2002a).

Gulmarg Wildlife Sanctuary consists of Subalpine Forests of Blue Pine *Pinus wallichiana*, Silver Birch *Betula utilis*, and Silver Fir *Abies pindrow*. Blue Pine forest is dominated by *Pinus wallichiana*, mixed at places with stands of Spruce *Picea smithiana*, Yew *Taxus wallichiana*, and Maple *Acer cappadocicum*. This type of forest occurs mainly at lower altitudes on dry aspects of the slopes (Bacha 2002a). Silver Fir is restricted to mountain folds and moist aspects. At lower levels, it is associated at some places with *Pinus wallichiana*, *Taxus wallichiana*, and *Picea smithiana*. Birch extends from 3,000 to 3,500 msl, and is distributed in the mountain folds and shady sites bordering alpine slopes. At lower altitudes, the forest has stands of Silver Fir (Bacha 2002a).

The alpine meadows have mainly herbaceous vegetation, with *Inula*, *Primula*, *Potentilla*, *Corydalis*, *Gentiana*, *Rumex*, and *Polygonum*. The meadows in and around Gulmarg were planted with many introduced bulbous species like iris, narcissus, daffodils, jonquils, and lupins, which have become naturalised.

Ibisbill *Ibidorhyncha struthersii* is threatened
by unregulated tourism in its habitat

ANANT ZANJALE

AVIFAUNA

There is no recent checklist of this site. According to Trevor Price (*pers. comm.* 2012) the bird list is likely to be similar to Overa-Aru and the upper reaches of Dachigam. The official list (Anon. 1987) needs to be confirmed. Osmaston (1923) recorded 76 species from Gulmarg area. Based on this old checklist, 14 species of Biome 7 (Sino-Himalayan Temperate Forest) and four of Biome 5 (Eurasian High Montane – Alpine and Tibetan) are found here. Among the threatened species, only Kashmir Flycatcher *Ficedula subrubra* is confirmed.

The Common Rosefinch is seen in large numbers in Gulmarg

Himalayan Snowcock *Tetraogallus himalayensis* is regularly seen. Himalayan Monal *Lophophorus impejanus* and Koklass Pheasant *Pucrasia macrolopha* have been recorded (Ifshan Dewan & M.S. Bacha, *pers. comm.* 2003).

OTHER KEY FAUNA

Large mammals recorded during a brief survey in 1979 include Rhesus Macaque *Macaca mulatta*, Brown Bear *Ursus arctos*, Himalayan Black Bear *Ursus thibetanus*, Red Fox *Vulpes vulpes*, Leopard *Panthera pardus*, and Himalayan Musk Deer *Moschus chrysogaster* (Green 1979, 1986). According to the Department of Wildlife Protection (Bacha 2002a) Markhor *Capra falconeri* is also found in this Sanctuary.

LAND USE
- ○ Tourism and recreation
- ○ Nature conservation and research
- ○ Pasture/Grazing land

THREATS AND CONSERVATION ISSUES
- ○ Unregulated tourism
- ○ Excessive livestock grazing
- ○ Cable car
- ○ Proposed amusement park
- ○ Collection of fuelwood

The Gulmarg tourist industry depends on the surrounding forest for fuelwood. The forests and lush green meadows, as in other protected areas of Jammu & Kashmir, suffer heavily from invasion of nomadic graziers during May to September.

KEY CONTRIBUTORS
M.S. Bacha, Ifshan Dewan, Khursheed Ahmad, and Intesar Suhail.

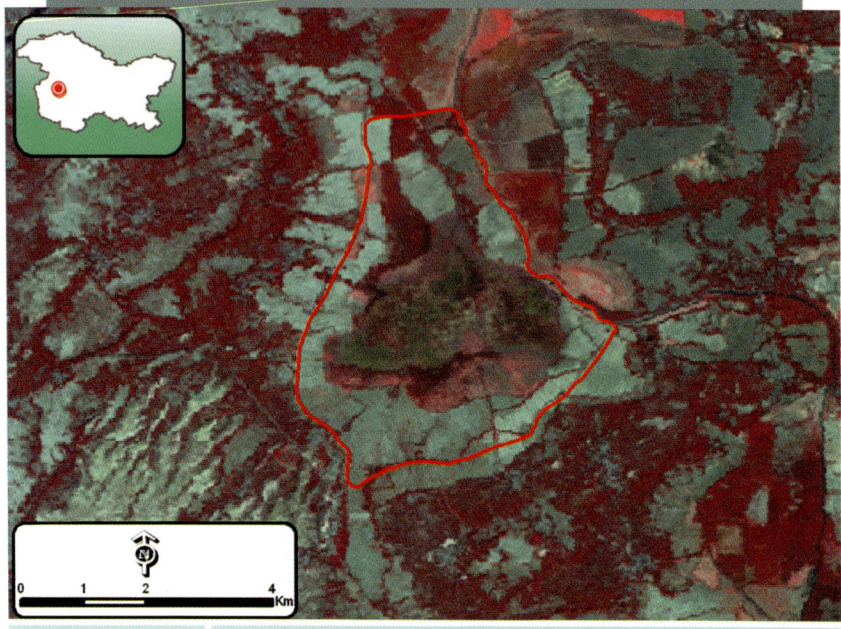

IBA Site Code	IN-JK-05
District	Baramulla
Coordinates	34° 17' 15" N, 74° 35' 46" E
Ownership	State Wildlife Department
Area	14 sq. km
Altitude	1,580 msl
Precipitation	900 mm + snowfall
Temperature	-2° C to 30° C
Biogeographic Zone	Himalaya
Habitats	Aquatic, Riverine Vegetation, Himalayan Secondary Scrub
IBA CRITERIA	: A1 (Threatened Species), A4iii (> 20,000 Waterbirds)
PROTECTION STATUS	: Wetland Conservation Reserve.

GENERAL DESCRIPTION

Haigam Rakh is named after village Haigam, and is situated *c.* 40 km from Srinagar. It was notified as a game reserve for duck shooting as far back as 1945. Earlier, the area was about 14 sq. km, with 4 sq. km of reedbeds (Holmes & Parr 1988) but now the total size of the reserve including the reedbeds has shrunk to 7.25 sq. km.

Haigam Rakh (marsh) is a permanent shallow freshwater lake with a maximum depth of 1.25 m, located in the Jhelum Valley at the southern end of Lake Wular. Perennial streams feed it, but the water table falls in late summer, reaches the lowest in autumn and rises again in early winter. Dissolved oxygen can reach very low levels in summer. The surrounding area is predominantly paddyfields and marshes with some fallow pastures that get flooded after heavy rain.

Most of the lake is covered with dense reeds and other emergent vegetation. Dominant species include *Typha angustata*, *Phragmites communis*, *Phalaris arundinacea*, *Sparganium erectum*, *S. ramosum*, *Scirpus lacustris*, and *S. palustris* (Kaul *et al*. 1980, Kaul 1982). In open areas there is floating vegetation such as Water Lilies *Nymphaea stellata* and *N. alba*, Fringed Water Lily *Nymphoides pellata*, and Water Chestnut *Trapa natans* (Kaul *et al*. 1980). The vegetation is rooted in the bottom of shallower areas or on floating mats of roots and silt (Holmes & Parr 1988). 183 species of phytoplankton have been recorded. Rows of willow have been planted recently on the edges of the lake, which is surrounded by paddyfields, orchards, and moist pasture.

AVIFAUNA

Haigam Rakh is the largest remaining reedbed area in the Kashmir Valley and is of major ornithological importance (Holmes & Parr 1988). It is particularly important for migratory species and marshland breeding species. Densities of the Little Bittern *Ixobrychus minutus*, Water Rail *Rallus aquaticus*, Common Kingfisher *Alcedo atthis* and the Clamorous Reed-warbler *Acrocephalus* [*stentoreus*] *brunnescens* are particularly high (Holmes & Parr 1988, Holmes & Hatchwell 1991). The area is important for autumn migrants, with 45% of the species recorded being passage migrants and/or winter visitors.

Haigam Lake is a major wintering area for migratory ducks, particularly Common Teal *Anas crecca*, Northern Pintail *A. acuta*, Eurasian Wigeon *A. penelope*, Mallard *A. platyrhynchos*, Gadwall *A. strepera*, Northern Shoveller *A. clypeata*, and Common Pochard *Aythya ferina*. It is also a vital breeding area for numerous species such as Little Grebe *Tachybaptus ruficollis*, Little Bittern *Ixobrychus minutus*, Little Egret *Egretta garzetta*, Water Rail *Rallus aquaticus*, Common Moorhen *Gallinula chloropus*, Pheasant-tailed Jacana *Hydrophasianus chirurgus*, and Whiskered Tern *Chlidonias hybridus*. Large numbers of hirundines and wagtails use the reedbeds for roosting and moulting. During the recent BNHS-MoEF Avian Influenza monitoring and surveillance study (2005–07) Khursheed Ahmad recorded a considerable population of White Wagtail *Motacilla alba* (794), Yellow Wagtail *M. flava* (794), Citrine Wagtail *M. citreola* (159), and Grey Wagtail *Motacilla cinerea* (40) here. The wetland is crucial for long distance migrants as a stopover site for feeding and resting.

Many waterbirds occur in huge numbers, much above the 1% population threshold determined by Wetlands International (2006). More recent records are available of surveys conducted by Khursheed Ahmad (2005–07) under the BNHS-MoEF Project on "Surveillance and Monitoring of Avian Influenza in Wintering Birds of India" (Rahmani *et al*. 2008). Based on published information (Scott 1989), the following species occur much above their 1% biogeographic population (total seen in Haigam : 1% threshold numbers): *Anas crecca* (7,000 : 4,000), *A. platyrhynchos* (25,000 : 750), *A. penelope* (3,000 : 2,500), *A. querquedula* (4,000 : 2,500) and *A. strepera* (4,000 : 1,500).

VULNERABLE	
Pallas's Fish-eagle	*Haliaeetus leucoryphus*

NEAR THREATENED	
Ferruginous Pochard	*Aythya nyroca*

Holmes & Parr (1988) also found that the very local Swinhoe's Reed-warbler *Acrocephalus concinens* (Blunt-winged Warbler of Grimmett *et al.* 1999) breeds in Haigam Rakh in small numbers, often near isolated willow trees. They found about 10 territories and fledged young ones in July-August 1983.

Historically, Bates & Lowther (1952) recorded breeding of the Ferruginous Duck *Aythya nyroca* in the smaller vales of Kashmir, particularly at Haigam, but Holmes & Parr (1988) found no evidence of breeding.

During the BNHS-MoEF Avian Influenza study (2005–07) Khursheed Ahmad recorded huge congregations of species such as Northern Pintail (9,524), Eurasian Wigeon (1,587), Mallard (9,937), Gadwall (7,937), Northern Shoveller (15,000), Common Teal (15,000), Red-crested Pochard (3,000) and Common Pochard (12,700) in Haigam Reserve. These were highest during February (Rahmani *et al.* 2008, Khursheed Ahmad, *unpubl.*). Khursheed Ahmad also reported sighting of Western Marsh Harrier *Circus aeruginosus* (8 individuals), Osprey *Pandion haliaetus* (2), and Peregrine Falcon *Falco peregrinus* (2) during the same study (Rahmani *et al.* 2008).

Haigam provides a vital staging area for many passage migrants including at least 18 species of shorebirds and several trans-Himalayan passage migrants. Pallas's Fish-eagle *Haliaeetus leucoryphus* has not been seen in the last 10 years (M.S. Bacha, *pers. comm.* 2003, Khursheed Ahmad *unpubl.* 2008), although Scott (1989) had reported upto five resident individuals.

INTESAR SUHAIL

Since shooting was stopped in 1995–96, there has been steady increase in the population of waterfowl

IMPORTANT BIRD AREAS OF J&K

Since shooting was stopped in 1995–96, a steady increase in waterfowl and other avian populations is seen at Haigam Rakh. The annual official waterfowl count indicates that total bird counts increased from a mere 1,530 birds recorded in 1996 to 169,305 birds in 1998, and 380,165 birds in 2002 (Bacha 2002c). In February 2012, a total of 350,000 birds were recorded in Haigam Wetland Reserve (Rauf Zargar, *pers. comm.* 2012) Even if these figures are excessive, there is no doubt that Haigam Rakh easily qualifies for A4iii criteria.

OTHER KEY FAUNA
Mammals include Common Otter *Lutra lutra* and Golden Jackal *Canis aureus*. The lake also supports a rich fish fauna.

LAND USE
○ Agriculture
○ Water management
○ Nature conservation and research
○ Encroachment by locals

THREATS AND CONSERVATION ISSUES
○ Siltation
○ Agricultural intensification and expansion in surrounding areas
○ Urbanisation

Haigam Rakh is protected as a conservation reserve by the Department of Wildlife Protection. Reed cutting is permitted, but the ban on hunting of waterfowl since 1995–96 has led to a marked increase in waterfowl populations.

The principal threats are siltation, encroachment due to agriculture, and urbanisation. Encroachers have now been evicted. Steps are being taken by the Wildlife Department to control eutrophication and weed infestation. Facilities are being built for tourists and birdwatchers. The State Government has asked the Central Government to include Haigam Rakh in the National Wetland Conservation Programme.

KEY CONTRIBUTORS
M.A. Parsa, M.S. Bacha, and Khursheed Ahmad.

HANLE PLAINS / MARSHES
WETLAND CONSERVATION RESERVE

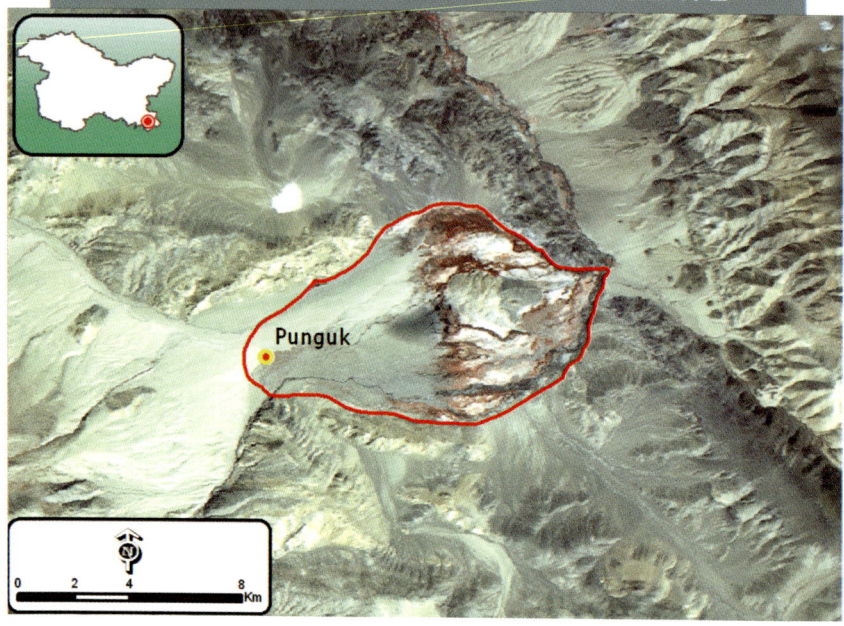

IBA Site Code	IN-JK-06
District	Leh, Ladakh
Coordinates	32° 47' 60" N, 79° 00' 00" E
Ownership	State / Private
Area	8,000 ha
Altitude	4,250–4,350 msl
Precipitation	100 mm + snowfall
Temperature	-40 °C to 30 °C
Biogeographic Zone	Trans-Himalaya
Habitats	Riverine Vegetation, Sandy Floodplains, Cold Desert Grasslands, Alpine Moist Pasture, Salty Marshes, Freshwater Pools
IBA CRITERIA	: A1 (Threatened Species)
PROTECTION STATUS	: Wetland is part of the Changthang Cold Desert Wildlife Sanctuary, also declared a Wetland Conservation Reserve by the Department of Wildlife Protection, J&K Government.

GENERAL DESCRIPTION

These marshes are located west and north of Hanle village in Ladakh, near the border with China. They are partly state owned and partly under the Hanle Buddhist monastery. The habitat is a complex of fast flowing streams, stagnant pools, saline marshes, seasonally flooded marshes, and bogs along the River Hanle, 45 km south of its confluence with the

Indus. The wetland marshes are frozen from December to April and are fed by snowmelt in summer.

The freshwater pools shelter vegetation such as *Hydrilla, Myriophyllum, Potamogeton*, and an edible aquatic lichen. Islam & Rahmani (2008) have suggested that Hanle Marshes be declared as a Ramsar Site as they fulfill Ramsar Criteria 2.

AVIFAUNA

The IBA is an important breeding ground for various waterfowl, including the Black-necked Crane *Grus nigricollis*, Ruddy Shelduck *Tadorna ferruginea*, Lesser Sand Plover *Charadrius mongolus*, and Redshank *Tringa totanus*. Three pairs of Black-necked Crane regularly breed in the Hanle Marshes and one in Lal Pathri about 22 km from Hanle (Chandan *et al.* 2006). During our recent surveys in July 2007, five breeding pairs of Black-necked Crane were recorded in the Hanle Marshes and adjoining Lal Pathri and Ramdo-Loma areas of Hanle. Out of these five pairs, two pairs with two chicks each were sighted at Tashicholin near Hanle Monastery and Lal Pathri Lagoon respectively (Khursheed Ahmad, *unpubl.* 2007). A pair of Redshank with a chick was also sighted in a marsh at Tashicholin. Subsequently, in June 2008, only one Black-necked Crane was sighted incubating an egg in a marshy pool at Lal Pathri area of Hanle (Khursheed Ahmad, *unpubl.* 2008). During autumn migration many birds pass through this site, including the globally Threatened Greater Spotted-eagle *Aquila clanga*. It uses the plain as the last staging

The marshes of the Hanle Plains are one of the most important nesting sites of Black-necked Crane

Excessive grazing by livestock is a threat to wild herbivores in the Hanle Plains

PANKAJ CHANDAN

VULNERABLE	
Greater Spotted-eagle	*Aquila clanga*
Black-necked Crane	*Grus nigricollis*

site before crossing the Himalayan range (Pfister 2001). Upland Buzzard *Buteo hemilasius* also breeds in this area.

OTHER KEY FAUNA

The slopes above Hanle Plains used to be an important habitat of Tibetan Gazelle *Procapra picticaudata*. In 1995–96 the first Tibetan Gazelle was seen here after 35 years of assumed local extinction. The sandy floodplains and riverine habitats of Hanle support a large population of Tibetan Wild Ass or Kiang *Equus kiang*. Tibetan Wolf *Canis lupus chanco* and Red Fox *Vulpes vulpes* are also found. Besides, Weasel *Mustela* sp., Blue Sheep or Bharal *Pseudois nayaur*, Argali *Ovis ammon*, Marmot *Marmota* sp., and Tibetan Woolly Hare *Lepus oiostolus* are commonly found.

LAND USE

- ○ Water management
- ○ Local recreation (no tourist permit is issued for this area)
- ○ Grazing land
- ○ High Altitude Astrophysics Station
- ○ ITBP posts

THREATS AND CONSERVATION ISSUES

- ○ Disturbance to birds from anthropogenic sources
- ○ Excessive livestock grazing competing with wild herbivores
- ○ Unplanned development and infrastructure
- ○ Plantation in the wetland
- ○ Wetland reclamation for agriculture
- ○ Diversion of water for agriculture

The area is used for grazing domestic livestock and for water supply to Hanle village. The human population in the valley is increasing, and correspondingly the livestock population, intensifying grazing pressure and use of water for high altitude agriculture. Packs of semi-feral dogs roam the region, taking a heavy toll of small mammals and nestlings, including unfledged Black-necked Crane. The dog numbers should be controlled, if the Black-necked Crane is to be saved.

Various parts of the marshes have been fenced for large-scale plantation of Willow under the supervision of the Forest Department. With the increasing influx of Tibetan refugees, their camps are growing bigger. Regular practice firing by the army and Indo-Tibetan Border Police disturbs the tranquillity of the site (A.K. Srivastava, *pers. comm.* 2000).

KEY CONTRIBUTORS

Otto Pfister, Rauf Zargar, Pankaj Chandan, and Khursheed Ahmad.

HEMIS HIGH ALTITUDE NATIONAL PARK

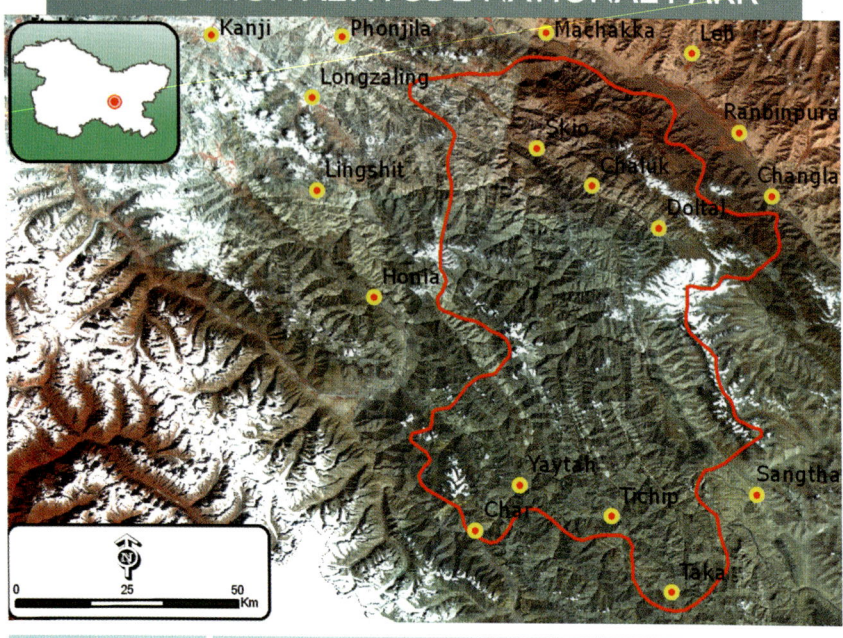

IBA Site Code	IN-JK-07
District	Ladakh
Coordinates	34° 01' 11" N, 77° 32' 00" E
Ownership	State Wildlife Department
Area	410,000 ha
Altitude	3,140–5,854 msl
Precipitation	110 mm + snowfall
Temperature	- 40 °C to 35 °C
Biogeographic Zone	Trans-Himalaya
Habitats	Alpine Dry Pasture, Riverine Vegetation, Alpine Moist Scrub, Rocky Cliffs

IBA CRITERIA	:	A3 (Biome 5: Eurasian High Montane)
PROTECTION STATUS	:	National Park, established in February 1981.

GENERAL DESCRIPTION

Hemis National Park is located in the Trans-Himalayan Ladakh district on the south bank of the Indus river. It extends from the southern side of the Indus Valley, southwards across the Zanskar Range as far as the Tsarap Chu and eastwards to the Buddhist Hemis Monastery, which the National Park is named after. The Markha and Rumbak Valleys and River Zanskar are located within the Park.

The present area of the Park is 410,000 ha, comprising a 125,000 ha core area and 285,000 ha buffer zone. Further extension of 65,000 ha has been recommended (Mallon & Bacha 1989).

Hemis is a wholly mountainous area. The core area (Alam Nullah and lower Chang Chu) lies in a band of hard limestone and other sediments that have been raised and tilted almost vertically, and deeply incised by a series of gorges. The terrain is extremely rugged with cliffs, screes, and exposed rock. It is isolated, with only a few passes crossing the main watersheds. The Markha and adjacent Sumdah Blocks comprise the catchment areas of the Markha, Rumbak, Shang, and Sumdah rivers, all of which drain north into the Indus. The area covered by these two blocks consists of narrow valleys and gorges. Gently sloping alluvial fans form a short section of stone desert along the south bank of the Indus between its confluences with the Zanskar and Rumbak. The upper Chang Chu (or Kharnak) Block lies above 4,000 msl. Here the landscape is different from the rest of the Park, and typical of the eastern plateau of Ladakh. This broad valley has a level floor upto 1 km wide, bounded by open hills with relatively few cliffs. Shun-Shadi, which encompasses the Niri Chu and Shun catchment, is a remote and sparsely inhabited block lying above 3,800 msl. Its terrain is exceptionally rugged, with deep gorges, cliffs, and steep, broken slopes. There are two small lakes, unusual features in the mountains of central Ladakh (Mallon & Bacha 1989).

Hemis is the largest protected area in the Indian Himalaya. Its large size and altitudinal range ensure that it is fully representative of the Trans-Himalayan ecosystem of central Ladakh. Important features are the remnant patches of Juniper scrub and riverine woodland, Snow Leopard *Uncia uncia* and associated prey populations, with an uninhabited and little-disturbed core area (Mallon & Bacha 1989). This PA has been selected as one of the several Snow Leopard Reserves under Project Snow Leopard by the Government of India, to conserve this rare species, its prey populations and its fragile montane habitat (Ministry of Environment and Forests 1988).

Much of central Ladakh is high altitude desert (Dhar & Kachroo 1983) characterised by sparse grassland and herbaceous vegetation on mountain slopes, with shrubs and patchy forest at the base of the valleys. The vegetation of the park is described by Mallon & Bacha (1989), and by Fox *et al.* (1986).

Trees are sparse, isolated, or in small open assemblies on hill slopes, and thin strips of riverine woods are common with the rest of Ladakh. The core area and the proposed Zanskar Gorge Block contain some of the best extant fragments of juniper forest steppe formerly common to many parts of Central Asia. Characteristic species are *Juniperus macropoda* and *J. indica*, which are scattered on cliffs and high slopes upto 4,250 msl, and form patches of open scrub in a few localities. Thin strips of riverine woodland are most extensive in the Chang Chu catchment. Principal species are *Salix karelinii* and *Myricaria squamosa*, with a few Poplar *Populus euphratica*, Silver Birch *Betula utilis*, Juniper *Juniperus* and Willow *Salix* spp. The vegetation thins out above 4,500 msl, with a few alpine species persisting to 5,000 msl and above. Chundawat (1990) gave a list of 314 plant species recorded in the catchment of Rumbak Nullah.

AVIFAUNA

Many high altitude birds of the Western Himalaya are found in Hemis NP, 80 species having been recorded till date, of which about 50 breed in the Park. The Park does not

Bearded Vulture *Cypaetus barbatus* is one of the largest birds of Hemis NP